HUGH HAMILTON

A

GEOMETRICAL TREATISE

OF THE

CONIC SECTIONS,

IN WHICH

THE **PROPERTIES** OF THE **SECTIONS** ARE DERIVED

FROM THE NATURE OF THE **CONE**,

IN AN EASY MANNER, AND BY A NEW METHOD

Elibron Classics
www.elibron.com

Elibron Classics series.

ISBN 1-4021-9821-3 (paperback)
ISBN 1-4021-2628-X (hardcover)

This Elibron Classics Replica Edition is an unabridged facsimile of the edition published in 1773 by J. Nourse, London.

A

GEOMETRICAL TREATISE

OF THE

CONIC SECTIONS.

A

GEOMETRICAL TREATISE

OF THE

CONIC SECTIONS,

IN WHICH

The PROPERTIES of the SECTIONS are derived from the Nature of the CONE,

In an EASY MANNER, and by a NEW METHOD.

By HUGH HAMILTON, A. M.

Fellow of TRINITY COLLEGE, DUBLIN, and of the ROYAL SOCIETY, LONDON, Now Dean of ARMAGH.

Tranflated from the LATIN ORIGINAL into ENGLISH.

LONDON,

Printed for J. NOURSE, in the Strand, Bookfeller to HIS MAJESTY.

MDCCLXXIII.

THE

TRANSLATOR's PREFACE.

THE following Treatiſe, written in Latin, and publiſhed in the year 1758, has been ſo well received by the learned, that the profeſſors in the ſeveral univerſities have uſed and recommended it in preference to all others on the ſame ſubject; and therefore the tranſlator thought that an Engliſh edition would be an acceptable preſent to the publick, in a country where ſo many, who had not much cultivated the learned languages, had yet diſtinguiſhed themſelves by their great proficiency in mathematical ſtudies.

As the author in his preface has given an account of the method in which he has deduced the properties of the Conic Sections, the tranſlator means only to add ſome obſervations here, which may ſhew wherein this method is preferable to any other that has been hitherto uſed.

Apollonius, and the other writers who have deduced the properties of the ſections from the cone, enquired but little further

further into the nature of the cone than was neceſſary to ſhew that three different ſections, beſides the triangle and circle, might be obtained by cutting its ſurface with a plane, and to demonſtrate in one particular inſtance that leading property of each Section, which is uſually called its equation. This was much too narrow a foundation on which to raiſe, with eaſe and elegance, ſo large a ſuperſtructure as the doctrine of Conic Sections. Some of the modern writers have betaken themſelves to a very different method, and founded this doctrine entirely on thoſe properties in each ſection, by which they may be deſcribed mechanically on a plane. But as theſe properties have a very remote relation to any of the general and fundamental properties of the ſections, their method is ſtill more prolix, as well as leſs natural, than the former one.

De la Hire, the famous French mathematician, who has written largely on this ſubject, deduces the properties of the ſections from the cone, and that he migth have more extenſive principles to proceed on, he has premiſed one entire book of lemmas, containing the properties of lines cut in harmonic proportion, particularly of ſuch as meet a circle, which is the baſe of a cone. Yet this expedient has rather embarraſſed the doctrine of Conic Sections, by making it depend ſo much on another that is really foreign to it. For there are but very few properties of theſe ſections that have any immediate connection with thoſe of lines cut harmonically; and ſuch the reader will find compriſed in a few propoſitions in our author's fifth book.

Now in all their reſearches on this ſubject, none of theſe learned writers ſeem to have ſuſpected that any of the general

neral properties of the fections, which they do not treat of till towards the conclufion of their works, might have been demonftrated in the beginning, and would have been an ample fource from whence to derive all the other properties. And yet nothing can appear more eafy and natural than the method of doing this, now that our author has difcovered it, and applied it fo fuccefsfully in the following Treatife. And, as we may collect from his preface, he feems to have been led to this difcovery by an obfervation, very ingenious indeed, but fuch as one would think might have been fufficiently obvious, viz. that whatever property agrees to all the fections in general, muft be a property of the conic furface on which they are placed. Such a property as this he has demonftrated of the cone in the eleventh propofition of his firft book, and deduced from it three other propofitions relative to the cone, which contain very material properties of the fections. He then proceeds to define the three conic fections, and the feveral lines belonging to them; and from the definitions he draws fome corollaries, to fhew that the properties before demonftrated of the cone may be equally affirmed of its fections.

Having thus laid down the principal properties of the fections in the beginning, he is enabled to deduce from thence the other properties with concifenefs and elegance, to arrange the feveral parts of his fubject in their moft natural order, and always to give general demonftrations for every property that is in itfelf a general one: by which means his work is freed from the intricacy and prolixity that have been complained of in other writers. We may therefore venture to fay that this work has juftly merited the

the applaufe it has met with; and doubtlefs it will continue to be admired while ever any tafte prevails for geometrical learning.

Nothing but an ardent defire of promoting this ufeful branch of fcience among his countrymen could have engaged the tranflator in this undertaking; and he flatters himfelf he has fucceeded fo far as to give the juft fenfe of the original, efpecially as he has been affifted in carefully revifing the whole by fome friends of acknowledged abilities.

London, March 21, 1773.

THE

AUTHOR's PREFACE.

IT is not necessary to say any thing here of the importance of that branch of geometry which treats of the properties of the Conic Sections. It has been very much cultivated by mathematicians both ancient and modern, and the learned well know its great use in astronomy, natural philosophy, and some of the practical arts.

Among the ancients who wrote on this subject, Apollonius was the most eminent, and on account of his excellent discoveries, and the accuracy of his demonstrations, obtained the name of the Great Geometrician. *Yet the learned and ingenious Doctor Wallis complains, that the doctrine of Conic Sections, as delivered by him, is perplexed and difficult, and that neither Mydorgius, nor others, had then removed this difficulty. Nor indeed do we find that any of the subsequent writers have treated this subject with sufficient clearness. The reason of which I take to be; that those among the moderns who defined the Conic Sections by their descriptions on a plane, were thereby obliged to deduce all their properties from a single one in each section, which is very remote from any of their general or fundamental properties. So that in this method many propositions must be*

 demon-

demonſtrated, and in a very operoſe manner, before thoſe fundamental properties can be treated of. The ancients indeed, and they who followed their method, judged rightly that the properties of the Sections were beſt derived from the Cone. And yet their works are difficult and prolix; becauſe they ſet about demonſtrating the properties of the Sections before they had ſufficiently conſidered the nature of the Cone. From whence it came to paſs that they demonſtrated with much labour many properties of each of the Sections ſeparately, which may, in a ſhort and clear manner, be proved to belong to the Cone itſelf, and may from thence be very eaſily transferred to all its Sections.

What has been here ſaid will, I believe, not appear ill founded to thoſe who compare the methods I have mentioned with that which is uſed in the following Treatiſe. What that method is, and the advantages ariſing from it, may be ſhewn in a few words.

In the elements of Geometry the properties of the triangle and circle (which are the two moſt ſimple of the Conic Sections) are eaſily deduced from their deſcriptions on a plane. But ſince the properties of the three others, which are to be the ſubject of the following Treatiſe, cannot be eaſily and conciſely deduced from their deſcriptions on a plane; I propoſe to retain their old definitions, and to explain the nature of the conic ſurface in ſuch a manner, that as ſoon as theſe curves are defined by the interſection of a plane with this ſurface, it may appear what their principal properties are. Accordingly, before the Sections are defined, certain properties of the conic ſurfaces are demonſtrated, as in Prop. XI. XII. XIII. XIV. Book I. *which contain in them moſt of the fundamental properties of the Sections. And then ſince it appears from their definitions, that theſe Sec-*

tion.

tions are curves, all the points of which are placed on a conic ſurface, it is manifeſt that every right line which any way meets theſe Sections, muſt in the ſame manner, and in the ſame points, meet the conic ſurface; and therefore all the properties which are proved to agree to right lines meeting the conic ſurface, are immediately transferred to thoſe which meet the Conic Sections. And thus the principal and moſt general properties of the Sections are laid down in the beginning, from whence we may reaſonably expect their particular properties will be more eaſily deduced.

Several advantages ariſe from treating the ſubject in this method. For thus every property which agrees to the three Sections may be always demonſtrated of them all jointly. Then by deducing particular properties from general ones the demonſtrations will be rendered ſhorter. Beſides, we ſhall have no occaſion to make uſe of Lemmas, or thoſe Propoſitions which are brought in only to demonſtrate others, with which works of this kind do too much abound; ſo that ſuch Propoſitions only will be inſerted as are of uſe in other ſciences, or ſeem to contain ſomething worthy of ſpeculation. By the help of this method, we ſhall be able to diſpoſe the Propoſitions in ſuch an order, that the properties, which are of the ſame kind, or relate to the ſame part of the ſubject, may be demonſtrated together. And laſtly, we ſhall perceive from hence the reaſon why theſe curves, which differ ſo much in ſome properties, agree entirely in others.

The demonſtrations of the theorems, and ſolutions of the problems, are moſtly new; as a new method and a new order of the Propoſitions neceſſarily require. And that the demonſtrations, like thoſe of the ancient geometricians, may be perfectly accurate, Evaneſcent quantities and points ſuppoſed to be removed to an infinite diftance, *which are now made ufe of by fome writers, find no place here.*

In the following Treatife I have inferted whatever I found in the works of others that feemed ufeful or proper for a work of this kind, and have introduced fome Propofitions, which, as far as I know, have not been delivered by others. Among which may be reckoned thofe before mentioned, which relate to the conic furfaces; as alfo Prop. LIV. Book I. Prop. XXIX. XXXVI. XXXVII. Book II. Prop. III. V. Book III. Prop. VI. IX. XI. XII. XIII. Book IV. Prop. XIV. Book V. *befides fome new Corollaries in different places.*

The reader will find, in fome places, references to fuch Propofitions of Sir Ifaac Newton's Mathematical Principles of Natural Philofophy, *as are more fully demonftrated in the following Work; which is chiefly defigned as an Introduction to the Newtonian Philofophy, the mathematical part of which is built on the doctrine of Conic Sections.*

Trinity College, Dublin,
March 20, 1758.

CONTENTS of the BOOKS.

BOOK I.

THE first nine Propositions contain such properties of the conic surfaces as have been delivered by others. Then certain properties are proved to belong to these surfaces, which are also principal and fundamental properties of the Sections, as appears afterwards from their definitions. The centers of the ellipse and hyperbola are found, and the properties which agree to all the diameters of the Conic Sections in general are demonstrated, and the proportion between the squares of their ordinates is shewn. The generation and properties of those lines that are called asymptotes of the hyperbola are clearly set forth, and the nature of conjugate hyperbolas explained. Several methods are shewn of drawing tangents to the Conic Sections. This Book also treats of the proportions between the squares and rectangles contained under the segments of lines which meet each other, and either touch or cut the Conic Sections; and also of the rectangles contained under lines drawn from points in the Sections to the sides of an inscribed trapezium, and which make given angles with those sides. It is proved that two Conic Sections cannot meet each other in more than four points.

BOOK II.

THE parameters of the Sections are defined, and those properties of the Sections demonstrated, from whence they have their denominations of ellipse, parabola, and hyperbola. The axes of the Sections are found,

found, and those of the ellipse and hyperbola are compared with their other diameters. The foci and directrices of the sections are defined, and the proportion is shewn between a line drawn from any point in a Section, and a perpendicular drawn from the same point to the directrix. These lines are proved to be equal when the Section is a parabola, and when from a point in an hyperbola a line is drawn to the focus, and another to the directrix, and parallel to the asymptote, these lines are also proved to be equal. It is proved, that if two lines be drawn from a point in the ellipse or hyperbola to the foci, their sum in the former, and difference in the latter, will be equal to the transverse axis of the Section: that a tangent to the parabola makes equal angles with two lines drawn from the point of contact, one to the focus, and the other a perpendicular to the directrix: and that a tangent to an ellipse or hyperbola makes equal angles with lines drawn from the point of contact to the foci. Various properties are demonstrated of lines passing through the foci of the Sections. A method is shewn of describing the Conic Sections on a plane by means of instruments. At the end of this Book it is demonstrated, that when a Conic Section is placed on the surface of a right Cone, if a plane passing through its directrix forms a circle on the Conic surface, the side of the Cone, intercepted between this circle, and any point in the Section, is equal to a line drawn from that point to the focus.

BOOK III.

THIS Book treats of the proportion of the parameters of the several diameters of a parabola, and of such properties of the parabola as are less analogous to those of the other sections; and also of the quadrature of the parabola, and the proportion of parabolic segments to each other.

BOOK

BOOK IV.

IN this Book such properties are treated of as belong only to the ellipse and hyperbola. The parallelograms circumscribed about their conjugate diameters, and about certain other diameters, are proved equal. It is demonstrated that in the ellipse the sum, and in the hyperbola the difference of the squares of any two conjugate diameters is equal to the sum or the difference respectively of the squares of their axes. The equal conjugate diameters of the ellipse are treated of. A new method is shewn for finding two conjugate diameters of the ellipse or hyperbola, which may contain given angles, by finding out the intersections of those curves with a circle, whose center lies in one of their axes. The intersection of an ellipse or hyperbola with a right line given in position is also found. The proportion of hyperbolic sectors and trapezia to each other is shewn, and the ellipse is compared with a circle described about its axis.

BOOK V.

THIS Book treats of Conic Sections that are similar to each other, and of such properties of the Sections as relate to lines cut in harmonic proportion. It is proved that the concourse of any two tangents of a Conic Section will lie in a right line given in position, when the line joining their points of contact passes through a given point. This Book likewise treats of many other general properties of the Conic Sections, and of the several points in which they may cut or touch each other: of the greatest and least lines that can be drawn from a point in the axis to the section: of circles that have the same curvature with the Conic Sections: and of the manner of describing Conic Sections, which may pass through given points, and touch right lines given in position.

OF THE

CONIC SECTIONS.

BOOK THE FIRST.

Of the Nature of the Cone, and of its Section.

DEFINITIONS.

I. IF through the point A, without the plane of the circle BDZC, a right line AZ be extended indefinitely both ways, and if (while the point A remains fixt) this line be carried round the whole periphery of the circle; each of the surfaces generated by this motion is called a *conical surface*, and both together are called *opposite surfaces*. FIG. 1.

II. The solid contained by the conical surface and the circle BDZ is called a *Cone*.

III. The point A is called the *vertex* of the cone.

IV. The circle BDZ, the *base of the Cone.*

V. A right line AZ passing through the vertex of the cone and any point in the periphery of the base, is called a *side of the Cone.*

A VI. The

VI. The right line AC, passing through the vertex and center C of the base and produced indefinitely, is called the *axis*.

VII. A *right Cone* is that whose axis is perpendicular to the base.

VIII. A *Scalene Cone* is that whose axis is inclined to the base.

IX. A right line meeting a conical surface in one point and produced both ways falling without the surface, is called a *Tangent*; but a right line meeting the surface, or the opposite surfaces in two points, is called a *Secant*.

Cor. A right line passing through any point in either surface and the vertex of the cone, is wholly in that surface; and being produced beyond the vertex, is in the opposite surface.

X. The common intersection of a plane with a conical surface is called a *Conic Section*.

PROPOSITION I.

Any plane passing through the vertex of a cone and cutting the periphery of the base, will cut the opposite surfaces in two right lines and in them only.

FOR if from the points, in which this plane meets the periphery of the base, two right lines be drawn to the vertex; they will be in this intersecting plane, and (by Cor. preceding) in the surface of the cone, and being produced beyond the vertex, in the opposite surface; that is, the plane will cut the opposite surfaces in these two right lines; and as the intersection of this plane with the plane of the base, cannot cut the periphery of the base in three points, it is evident that the plane cannot cut the opposite surfaces in three right lines; therefore the plane will cut the surfaces in two right lines only.

Cor. The common intersection of any plane drawn through the vertex with the conical surface and base, will be a triangle.

PROP.

PROP. II.

A right line ED joining any two points E, D, in a conical surface, provided they be not in the same right line passing through the vertex, falls wholly within the surface, but produced both ways without it, and does not meet again either of the surfaces. On the contrary, a right line ED joining two points E, D, in the opposite surfaces, provided they be not in the same right line passing through the vertex, falls without both surfaces, and produced both ways it falls within them, and meets neither of them again. Fig. 2. Fig. 3.

FOR the sides AEC and ADB of the cone being drawn through the points E and D, a plane passing through these lines does not meet the surface or opposite surfaces, but in the lines AEC, ADB; therefore the line DE does not meet them, but in the points D, E.

In the first case the plane of the angle DAE and consequently the line DE, is within the conical surface; and therefore produced both ways falls without it. Fig. 2.

In the second case the plane of the angle DAE and consequently the line DE is without both the opposite surfaces; and therefore DE produced both ways falls within the opposite surfaces. Fig. 3.

If the points D, E, be in the same right line passing through the vertex, the line joining these points (Cor. to Definitions) will be wholly in the conical surfaces.

PROP. III.

If a right line DE touches the periphery of the base of the cone in the point D, and from A the vertex of the cone, the right line AD be drawn to the point of contact; the plane ADE passing through both these lines Fig. 4.

 touches

touches the conical ſurface in the line AD only. The meeting of this plane with the ſurface is called the line of *Contact.*

FOR the line AD is in the conical ſurface, and in the plane ADE, and as every other point beſides D, as F in the periphery of the baſe, is without the line DE; it follows that every other line beſides AD, as FA in the conical ſurface, is without the plane ADE, and hence the propoſition is manifeſt.

Cor. 1. Hence, and from Defin. I. it appears that the plane ADE produced touches the oppoſite ſurfaces in the line AD produced.

Cor. 2. Hence appears a method of drawing a plane which ſhall touch a conical ſurface in a given right line, as AD; for the right line DE being drawn in the plane of the baſe, touching the periphery in D, the plane ADE drawn through AD, DE touches the ſurface in the line AD.

Cor. 3. If any right line ED touches the conical ſurface in the point D, and AD be drawn from the vertex; the plane paſſing through AD, DE touches the ſurface. For through the vertex A in the plane ADE, draw AG parallel to ED; this right line meets all the ſides of the cone in the vertex; therefore if any other ſide of the cone beſides AD were in the plane ADE, the line ED parallel to AG would neceſſarily meet that ſide, and then ED would meet the conical ſurface in two ſides, contrary to the hypotheſis; wherefore the plane ADE does not meet the conical ſurface, but in the line AD.

Cor. 4. No other plane beſides ADE touches the conical ſurface in the line AD; for its interſection with the plane of the baſe will cut its periphery. Hence it follows that this plane cuts the conical ſurface in two right lines (by Prop. I.) and therefore does not touch it.

PROP. IV. PROBL. I.

Fig. 5, 6. **Through a right line AC, paſſing through the vertex of the cone, but not in its ſurface, to draw a plane which ſhall touch the conical ſurface.**

THROUGH

THROUGH the line AC let any plane be drawn, cutting the base in GH, which intersection the line AC (being in the same plane) will either meet, or be parallel to: first, let them meet in E; and from this point on either side of EHG let a right line be drawn touching the periphery of the base; and join AD, and ADE will be the plane required; for it passes through AC, and (by the preced.) touches the surface. Secondly, let AC be parallel to GH; and let a diameter MD of the base be drawn perpendicular to GH and meeting the periphery in D, and through the point D draw DE touching the base; it will be parallel to GH and consequently to AC; then joining AD, ADE will be the plane required; for because AC and DE are parallel, AC will be in the plane ADE and this plane touches the conical surface (by the preced.) FIG. 5. FIG. 6.

COR. It is evident that only two planes passing through AC can touch the conical surface, viz. one on each side of the plane AGH.

PROP. V.

Any right line ED parallel to a side AB of the cone (provided it be not in the plane touching the surface in AB) meets one of the opposite surfaces and that in one point E only; that is, on one side it is wholly without both surfaces, and on the other wholly within that where the point E is taken. FIG. 7.

LET a plane passing through the parallels AB, ED, cut the surfaces in *c*AC; this plane does not meet the surfaces but in the lines AB and *c*AC; and the line ED in this plane does not meet AB (for they are parallel) but necessarily meets *c*AC somewhere, as in the point E, and on one side of the point E is within the angle BAC, and on the other part within the angle adjacent to it. Whence the proposition is evident.

COR. It appears that the plane passing through DE and parallel to the plane touching the conical surface in the line AB, does not meet the opposite surface.

PROP.

PROP. VI.

Fig. 8. **Every right line EF parallel to a right line AD, paſſing through the vertex A, and falling within the conical ſurfaces, meets each ſurface once as in the points E, F; and**
Fig. 9. **every right line EF drawn through any point P within either ſurface, and parallel to any right line AD paſſing through the vertex and falling without the ſurfaces, meets the ſame ſurface ſomewhere, as in the points E, F.**

LET the plane paſſing through AD and EF cut the conical ſurfaces in the lines BA*b*, CA*c*, which it is manifeſt may be always done; ſince in the firſt caſe AD falls within the angle BAC, the line EF, which is parallel to AD, and in the ſame plane with the angle BAC, neceſſarily meets the ſides AC, AB, viz. AC in E, and AB produced in the point F of the oppoſite ſurface: and in the ſecond caſe, ſince AD falls without the angle BAC, the line EF parallel to it, drawn through the point P within this angle, neceſſarily meets AC, AB ſomewhere [illegible] is, meets the ſurface in the points E, F, within which the point [illegible].

Cor. 1. Hence, and from Prop. II. it appears that a right line does not meet a conical ſurface or oppoſite ſurfaces in three points.

Cor. 2. If a right line be drawn parallel to any right line EF meeting the oppoſite ſurfaces; this line will likewiſe meet the oppoſite ſurfaces. For if AD be drawn through the vertex A parallel to EF it is evident that it falls within the conical ſurfaces; therefore any right line drawn parallel to EF will be parallel to a right line AD paſſing through the vertex and falling within the ſurfaces, and conſequently it meets the oppoſite ſurfaces.

Cor. 3. Since the line EF joins two points in the ſame ſurface, the line AD drawn through the vertex parallel to it falls without the ſurface; for it is manifeſt that it falls without the angle BAC, the ſides of which EF meets.

Cor.

COR. 4. Every plane passing through this line EF, in the first case, will cut both surfaces; and in the second case, if a plane falling without the conical surfaces, passes through AD; the plane drawn through EPF parallel to it, will cut the conical surface within which the point P is taken throughout its whole circumference, and will not meet the opposite surface.

PROP. VII.

If either of the opposite surfaces be cut by a plane GHF parallel to the base, the section EGFL will be the circumference of a circle whose center is in the axis of the cone. FIG. 10.

LET C be the center of the base and AC the axis of the cone, and let the cone be cut in any manner through the axis by two planes making the triangles ABD, AIK, which (produced if necessary) may meet the plane GHF in the lines GHL, EHF; the planes being parallel, the triangles AHF, ACK, and likewise AHL, ACD, are similar: wherefore HL is to CD as AH to AC, and likewise HF to CK will be in the same ratio; whence by permutation, HL is to HF as CD to CK: but CD and CK are equal; therefore HL and HF are equal. In like manner it may be demonstrated, that any other two lines, as HG, HE, drawn from the point H to the section EGFL, are equal; therefore that section is the circumference of a circle (by 9. 3.) and its center H is in the axis of the cone as is evident.

COR. 1. Hence it appears that if a diameter of such a section be drawn, it will cut the axis of the cone: and conversely, if a right line drawn in the plane of this section cuts the axis of the cone, it will be a diameter of the section.

COR. 2. It likewise appears that any circle parallel to the base may be taken for the base of the cone.

PROP.

PROP. VIII.

Fig. 11. If a scalene cone ABCL be cut through the axis by a plane perpendicular to the base making the triangle ABL, and from the vertical angle BAL another triangle ADI be cut similar to ABL, but in a subcontrary position, that is, so that the angle ATD be equal to the angle ABL; and if the cone be cut again by a plane passing through DT and perpendicular to the plane of the triangle ABL: the section DKTF in the conical surface will be the circumference of a circle.

LET the right line MN be the common intersection of the plane DKTF with the plane of the base: from any point K, in the section DKTF, draw KGF parallel to MN, and meeting DT in G, through G draw EH in the plane ABL and parallel to BL; then the plane passing through KGF, EGH will be parallel to the base (15. 11.) and therefore perpendicular to the plane ABL, and its intersection with the surface of the cone will be a circle, of which EGH is a diameter (by the preced. and Cor. 1.) and because both these planes are perpendicular to the plane of the triangle ABL, their common intersection KGF will be perpendicular to the plane ABL (19. 11.) and therefore to both these right lines EH, DT; wherefore the square of KG (by a property of the circle) is equal to the rectangle EGH; but the angles ATD and (ABL or) AEH are equal (by hypothesis) and the vertical angles at G are equal; therefore the remaining angles EDG, GHT are equal, and therefore the triangles EDG, GHT are similar; wherefore DG is to EG as GH to GT; and therefore the rectangle DGT is equal to the rectangle EGH, that is, to the square of KG. Therefore the point K is in the circumference of the circle described about the diameter DT in the plane DGK: for if it were not, the rectangle DGT would be equal to the square of a line greater or less than KG, contrary to what has been proved: in like manner the same may be shewn of any other point in the section DKTF, which is therefore the circumference of a circle. This section is called a *subcontrary section.*

PROP.

PROP. IX.

If in the conical surface the section DKTF made by a plane not parallel to the base be the periphery of a circle, it will be a subcontrary section. FIG. 11.

LET a conical surface be cut by two planes parallel to the base making the circles EKHF, OPQR and cutting the plane DKTF in the lines KF, PR, which will be parallel (by 16. 11.) let the diameters EH, OQ of these circles be drawn perpendicular to the lines KF, PR and meet them in the points G, S; the lines EH, OQ will be parallel (Converſ. 10. 11.) and will bisect the lines KF, PR in G and S (3. 3.) and will pass through the axis of the cone (by 1. Cor. 7. of this Book) therefore the plane AOEBLHQ passes through the axis of the cone; let the line DT be its intersection with the plane of the circle DKTF; since DT passes through the points G and S and consequently bisects the lines KF, PR terminated by the circle DKTF, it is a diameter of that circle and is perpendicular to these lines as appears from the (3. 3.) Then because the line KF is perpendicular to the lines DGT, EGH, it will be perpendicular to the plane ABL passing through these lines; therefore (by 18. 11.) the planes of the circles DKTF, EKHF and consequently the plane of the base will be perpendicular to the plane ABL passing through the axis of the cone; but (by a property of the circle) each of the rectangles EGH, DGT is equal to the square of KG and consequently are equal to each other; therefore DG will be to EG as GH to GT, and the vertical angles at G are equal; therefore the triangles DGE, HGT (by 6. 6.) are similar, and therefore the angle ATD is equal to the angle AEH, that is to the angle ABL, wherefore by the preced. Prop. the section DKTF is a subcontrary section.

COR. Hence a conic section, which is not a subcontrary section, nor parallel to the base, will not be the circumference of a circle.

PROP. X.

FIG. 12, 13. Let two right lines meet each other in the point P, and let one of them be parallel to the base of a cone, and the other parallel to a right line given in position cutting the plane of the base: according as both touch or cut, or one of them touches, and the other cuts a conical surface, or opposite surfaces; the squares of the segments of the tangents, or the rectangles under the segments of the secants, between their point of concourse, and the points where they meet the conical surface or surfaces, will be in a constant ratio to each other, wheresoever P the point of concourse of these two lines is taken. *I call a right line parallel to a plane, when in that plane a right line can be drawn parallel to it.*

LET the circle ASC be the base of the cone ABC, and BM the right line given in position drawn from the vertex B to any given point M in the plane of the base (but not in its periphery) and from the point M let any right line be drawn cutting the periphery of the base in the points K, L; let two right lines meet each other in any point P, and let one of them POG or PT be parallel to the base of the cone and cut the surface in the points O, G, or touch it in T, and let the other line PEF be parallel to BM; first let this line PEF cut the same surface or opposite surfaces in E, F; the rectangle EPF will be to the rectangle OPG or to the square of PT, as the square of BM to the rectangle KML, wheresoever P the point of concourse of the right lines may be taken.

For through the line POG let a plane be drawn parallel to the base making the circle DOHG, and through the parallels BM and PEF let another plane pass meeting the surfaces in the lines QEA, RFC, the plane of the base in the line MCA, and the plane of the circle DOHG in the line DPH; MCA, DPH are parallel (16. 11.) therefore the triangles EPD, BMA, are similar, and also the triangles

TAB·I·
Page·10·

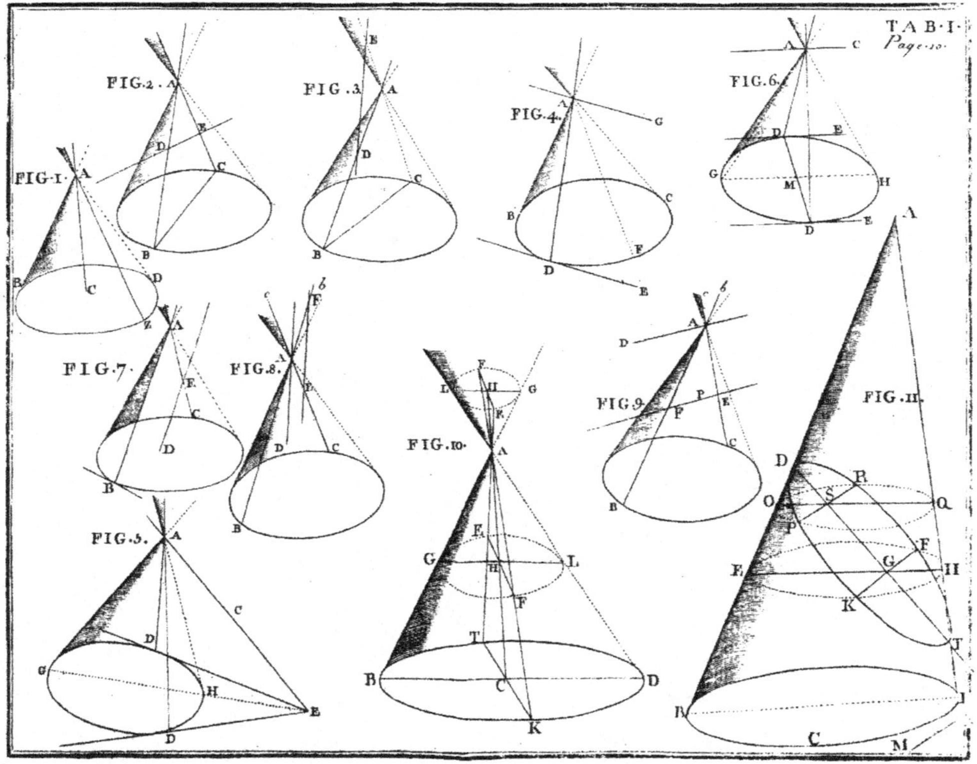

angles FPH, BMC, and consequently EP is to DP as BM to AM, and PF to PH as BM to MC; therefore (by taking the rectangles under the antecedents of one order of proportionals and the antecedents of the other, and the rectangles under the consequents of one, and the consequents of the other) the rectangle EPF is to the rectangle DPH as the square of BM to the rectangle AMC by (23. 6.) but because DOHG and ASC are circles, the rectangle DPH is equal to the rectangle OPG or to the square of PT, and the rectangle AMC is equal to the rectangle KML by (35 and 36. 3.) therefore the rectangle EPF is to the rectangle OPG or the square of PT, as the square of BM to the rectangle KML, wheresoever P the point of concourse of the lines EF, OG is taken.

But if the line EF parallel to BM meets in P the line PB parallel to the base of the cone and passing through the vertex; then in like manner it may be demonstrated (from the similar triangles EPB, BMA; and FPB, BMC) that the rectangle EPF is to the square of PB as the square of BM to the rectangle AMC, that is, to KML.

In the second place, the rest remaining as before, let the right line PF touch the surface in F, and through the point of contact F let the side of the cone be drawn meeting the base in N; the plane passing through the parallels PF, BM touches the surface in the line BFN (3. Cor. 3. of this Book) and its intersection MN with the plane of the base, touches the base in N; let the line PT be the intersection of the same plane with the plane of the circle OTG in which the point P is; this line touches the circle OTG in the point T, viz. in the line BFN, and will be parallel to MN (16. 11.) therefore, from the similar triangles FPT, BMN, the square of PF will be to the square of PT or to the rectangle OPG, as the square of BM to the square of MN, or to the rectangle KML. Q. E. D. FIG. 14.

PROP. XI.

If two right lines meeting each other be parallel to two right lines given in poſition; according as both touch or cut, or one of them touches, and the other cuts a conical ſurface or oppoſite ſurfaces; the ſquares of the tangents, or the rectangles under the ſegments of the ſecants between their point of concourſe and the points wherein they meet the ſurface or oppoſite ſurfaces, will be in the ſame ratio to each other whereſoever the concourſe of the two right lines falls.

Caſe 1. WHEN both the lines are parallel to the baſe of the cone, a plane paſſing through them will be parallel to the baſe (15. 11.) and therefore will cut one of the ſurfaces in the circumference of a circle; and therefore both the lines will either touch or cut, or one of them will touch and the other cut the ſame circle; and therefore the ſquares of the ſegments of the tangents, or the rectangles under the ſegments of the ſecants between the point of concourſe of the two lines and the points wherein they meet the ſurface will be to each other in a ratio of equality by the 35 and 36. 3.

Caſe 2. When one of the lines meeting each other is parallel to a line given in poſition cutting the plane of the baſe, and the other parallel to a line given in poſition drawn in the plane of the baſe; this line will be parallel to the plane of the baſe, and therefore this caſe is manifeſt from the preceding Propoſition.

Caſe 3. When neither of the lines meeting each other is parallel to the plane of the baſe; through their point of concourſe let a right line be drawn parallel to the plane of the baſe, and cutting either ſurface, or perhaps paſſing through the vertex of the cone; then according as either of the lines not parallel to the baſe touch or cut the ſurface or oppoſite ſurfaces, the ſquare of the ſegment or the rectangle under the ſegments of the ſecant, will be in a conſtant ratio to the ſquare of the ſegment, or to the rectangle under the ſegments of the line parallel to the plane of the baſe (by preceding Prop.) and

and consequently these squares or rectangles under the segments of the lines which are not parallel to the plane of the base, will be in a constant ratio to each other as appears from (22. 5.) *

In this and the preceding proposition when one of the right lines which meet each other passes through the vertex of the cone, the square of the segment between the point of concourse and the vertex, must be taken as if the right line was a tangent, although it may happen to fall within the conical surface.

PROP. XII.

If two parallel right lines meet a right line parallel to a side of a cone; according as both touch or cut, or one of them touches, and the other cuts a conical surface or opposite surfaces; the squares of the segments of the tangents, or the rectangles under the segments of the secants between the surface or opposite surfaces, and the right line which they meet, will be to each other as the segments of that right line between the parallels, and the point wherein it meets the conical surface.

LET MBC be a cone, and the right lines EF, LN parallel to each other, and let them cut in any manner the same surface or opposite surfaces in E, F and L, N, and in the points P, Q meet the line ST parallel to *c*BC the side of the cone, and meeting the surface in V; the rectangle EPF will be to the rectangle LQN as PV to QV. FIG. 15.

For let the plane passing through the parallels *c*BC and ST meet again the surfaces in the line *m*BM, and through the points P, Q draw in this plane the lines MPC, DQH parallel to each other, and meeting the surface in M, C and D, H; then (by the preceding) the rectangle DPF is to the rectangle MPC as the rectangle LQN to

* The 22d Proposition of the fifth book of Euclid here referred to is that by which the way of reasoning in proportions called, *ex æquo ordinatim* is proved, and from thence the above conclusion will follow. It is also the Proposition by which Euclid proves the eighth of his *Data*, which is, that two quantities which have a given ratio to a third, have also a given ratio to each other; therefore the eighth of Euclid's *Data* might have been more properly referred to on this occasion.

the

the rectangle DQH; therefore by permutation the rectangle EPF is to LQN as the rectangle MPC to DQH, that is, becaufe PC and QH are equal (being oppofite fides of a parallelogram) as MP is to DQ, or from the fimilarity of triangles as PV to QV; the fame may be fhewn concerning the fquares of the fegments of the parallels between the line ST and the point of contact, if one or both be tangents. Q. E. D.

PROP. XIII.

Let two parallel lines touch a furface, or oppofite furfaces, and two points be taken in them equidiftant from their points of contact; if the right line joining thefe points meet the furface or oppofite furfaces; its fegments between the tangents and furface or furfaces, will be equal.

Fig. 16. LET the right lines EP, RQ be parallel, and touch the fame furface or oppofite furfaces in E and R, and let the points P, Q be equidiftant from the points E, R, and let PQ be drawn cutting the furface or oppofite furfaces in the points F, G; the fegments PF, QG will be equal.

For becaufe the lines EP, RQ are parallel, and the fame right line PQ meets them; the fquare of EP will be to the rectangle FPG as the fquare of RQ to the rectangle GQF (by the 11. of this Book) but the fquares of EP, RQ are equal (by hypothefis) therefore the rectangles FPG and GQF are equal; let FG be bifected in O; and when the point F, G are in the fame furface, let the equal fquares of OF and OG be added to the equal rectangles FPG, GQF and the fums that is the fquares of OP, OQ (by 6. 2.) are equal; and when the points F, G are in oppofite furfaces, let the equal rectangles FPG, GQF be taken from the equal fquares of OF, OG and the remainders, that is (by 5. 2.) the fquares of OP, OQ will be equal; therefore the right lines PO, OQ in both cafes are equal; but FO, GO are equal, therefore the fegments PF, QG will be equal.

But

But if the line PQ touches the surface in the point V; then, as before, (by 11. of this Book) the square of EP is to the square of VP as the square of RQ to the square of VQ; therefore the squares of VP, VQ, and consequently the segments VP, VQ are equal. Q. E. D.

Cor. Let the parallel lines PF, QH touch the conical surface in the points E, R, draw ER, and through the point O taken in it draw a right line parallel to PF, QH meeting the surface in the points M, N: if through the points M, N two lines be drawn parallel to ER and meeting PF, QH in the points F, H and P, Q; these lines either at the same time cut the surface again in the points G, L, or at the same time touch it in the points M, N. By construction and (34. 1.) the tangents EF, RH and likewise EP, RQ are equal; then if the line FMH meets the surface in the points M, G, FM, GH will be equal (by this Prop.) and if the line PNQ be supposed to touch the surface in N, PN, NQ will be equal; therefore, because PF, NM, QH, are parallel, FM, MH will be equal, which is absurd; therefore when the line FMH meets the surface in two points, PNQ will not touch the surface in N; consequently it meets it again; wherefore FMH, PNQ will either at the same time cut the conical surface, or at the same time touch it. Fig. 17.

PROP. XIV.

The right line joining the two points of contact of parallel tangents bisects all right lines which it meets parallel to those tangents, and terminated both ways by either of the surfaces.

LET the lines EP, RQ be parallel, and touch the same surface or opposite surfaces, and draw ER joining the points of contact; then let the right line MN be parallel to EP, RQ, and meet ER (produced if necessary) in the point O, and let it be terminated Fig. 17, 18.

nated by the conical surface in the points M, N; the segments MO, ON will be equal. For draw through M, N two right lines parallel to ER, and first, both meeting (by Cor. preced.) the same or opposite surfaces again in G, L, and the parallel tangents in the points F, H and P, Q; then the lines MF, NP, being opposite sides of a parallelogram, will be equal, and likewise FH, PQ and the tangents EP, RQ; therefore (by the preced.) the lines NP, QL will be equal: and in like manner it may be shewn that MF, HG are equal; therefore PL, FG will be equal and consequently the rectangles NPL, MFG are equal; but (by 11 of this Book) the rectangle NPL is to the square of PE as the rectangle MFG to the square of EF; therefore the squares of EP and EF are equal and consequently the lines EP, EF, and therefore the segments MO, ON which are parallel to them, will be equal.

FIG. 18. When the points E, R are in opposite surfaces, the lines drawn through M, N parallel to ER necessarily meet the opposite surfaces (by 2. Cor. 6. of this Book) but when the points E, R are in the same surface, if one of the lines which are drawn through M, N parallel to ER, touches this surface in the point M, the other line touches it in N (by Cor. preced.) let these lines meet the tangent EPF in the points X, V; then the tangents MX, NV, being opposite sides of a parallelogram, will be equal; therefore (by the preced.) the tangents XE, EV will be equal, and therefore the segments MO, ON being parallel to them, will be equal. Q. E. D.

DEFINITIONS.

FIG. 19. XI. If a plane ADE touches a conical surface in the right line AD, and the surface be cut by a plane parallel to ADE; the section MVN is called a *Parabola.*

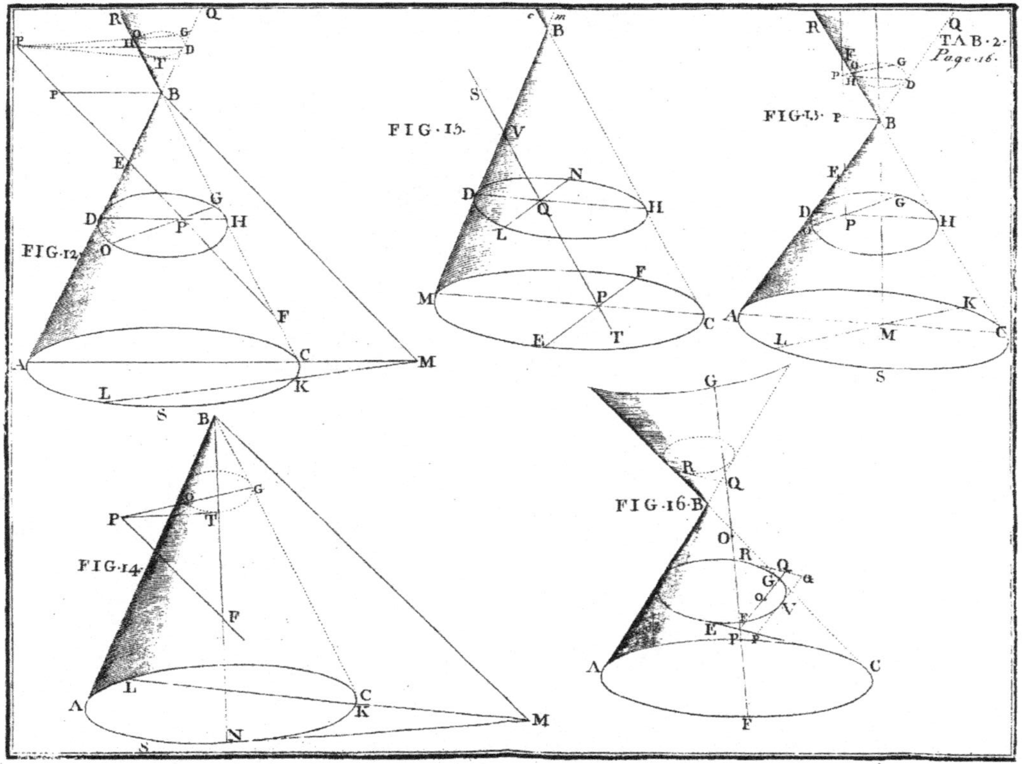
TAB·2·
Page·16·
FIG·12·
FIG·13·
FIG·14·
FIG·15·
FIG·16·

XII. If through the vertex and without the conical surfaces a plane DAE passes, which is neither parallel to the plane of the base nor to the plane of subcontrary section, and if either surface be cut by a plane XVLT parallel to DAE; the section MVNL is called an *Ellipse*. FIG. 20.

XIII. If through A the vertex of the cone BAG a plane passes cutting the surfaces in the right lines *d*AD, *e*AE, and the conical surface be cut by a plane XVM parallel to EAD; the section MVN is called an *Hyperbola*. And because this intersecting plane necessarily meets the opposite surface, and there makes a section *m*L*n* of the same kind and denomination with the former, these two sections together are called *opposite sections*. FIG. 21.

XIV. The plane ADE passing through the vertex of the cone and parallel to the plane of any section, I call the *vertical plane of that section*.

XV. Any right line LV drawn in the plane of a parabola, and parallel to the right line AD in which the vertical plane touches the surface, is called a *diameter* of the parabola, and the point V where it meets the section is called its *vertex*. FIG. 19.

XVI. The point within an ellipse, or between the opposite sections, in which is bisected every right line drawn through it, and terminated both ways by the ellipse or by the opposite sections, I call the *center* of the ellipse, or of the hyperbola, or of the opposite sections.

XVII. Any right line passing through the center of an ellipse or hyperbola, and meeting the ellipse or opposite sections in two points, I call a *diameter* of the ellipse, or a *transverse diameter* of the hyperbola or of the opposite sections; and the points in which the diameter meets the ellipse, or opposite sections, are called its *vertices*.

XVIII. Any right line passing through the center of an hyperbola, and bisecting a right line not passing through the center and terminated by the opposite sections, is called a *second diameter* of the hyperbolas.

XIX. Two diameters, each of which bisects all right lines terminated by the ellipse, hyperbola, or opposite hyperbolas and parallel

rallel to the other diameter, are called ***conjugate diameters*** of the ellipse or hyperbola.

XX. Any right line, not passing through the center, terminated both ways by a conic section or the opposite sections, and bisected by a diameter, is called an ***ordinate applied*** to that diameter, or more simply an ***ordinate:*** and more frequently the half of this right line is called an ***ordinate***.

XXI. The segments of a diameter between its ordinate and vertices are called *abscisses*.

XXII. A diameter of a conic section, which cuts the ordinates at right angles, is called an *axis*.

XXIII. A right line drawn in the plane of a conic section, and meeting the section in one point, and produced both ways falling without it, is said to ***touch*** the section in this point, and this line is called a ***tangent***; but if a right line meets the section or opposite sections in two points, it is called a *secant*.

Corollaries to the preceding Definitions.

Cor. 1. As every point in a conic section is in the conical surface, and any two points in the opposite sections are in the opposite surfaces, it is evident that every right line drawn in the plane of a conic section, and meeting in any manner the conical surface or opposite surfaces, will in the same manner and in the same point or the same points, meet the conic section or opposite sections: and conversely it is evident that every right line meeting in any manner a conic section or opposite sections, will in the same manner and in the same point or the same points, meet the conical surface or opposite surfaces.

Cor. 2. A right line joining any two points in a conic section falls wholly within the section, and produced both ways falls without it, and does not meet either it or the opposite section again; and conversely a right line joining two points in the opposite sections falls without both sections, but produced both ways falls within each section and

meets

meets neither of them again. This corollary is manifest from the preceding and (Prop. II. of this Book). Hence or from (Cor. 1. 6. of this Book) it appears that a right line cannot meet a conic section or opposite sections in three points.

Cor. 3. Hence the same right line cannot touch the opposite sections.

Cor. 4. Hence likewise if in the plane of the opposite sections a right line be drawn parallel to the line joining two points in the opposite sections, it will necessarily meet the opposite sections, as appears from Cor. 2. Prop. VI.

Cor. 5. All the diameters of a parabola are parallel to each other, and any line drawn in the plane of a parabola and parallel to the diameters, will be a diameter as appears from Definition 15. and each diameter meets the parabola in one point only and is on one side wholly without the parabola, and on the other within it; as appears from Prop. V. and Cor. 1. to the Definitions.

Cor. 6. It is manifest since the planes of the parabola and hyperbola cut the conical surface (though produced indefinitely) that these sections may be extended *ad infinitum*, and never return again or include a space; but the ellipse surrounds the cone, and returning into itself includes a space.

Cor. 7. Every conic section surrounding the cone is either a circle or an ellipse; for if the plane of the section be parallel to the base, or in subcontrary position, it will be a circle, but if not, it will be an ellipse by Definition 12. and Cor. Prop. IX. of this Book.

PROP. XV.

If through a point within a parabola a right line be drawn meeting the diameters, it neceſſarily meets the parabola in two points.

Fig. 19. LET BAD be a cone, and the ſection MVN in its ſurface a parabola, and the plane ADE its vertical plane touching the ſurface in the line AD, and let LV be a diameter; and through the point P within the parabola draw the right line MN cutting the diameters, it will meet the parabola in two points.

For draw in the vertical plane, through the vertex A of the cone, the right line AF parallel to MN; then becauſe the lines PN, FA, and PL, AD are parallel, the angles NPL, FAD will be equal (10. 11.) therefore AF will cut the line AD in which line only the vertical plane meets the ſurface: therefore AF falls wholly without the ſurface: and therefore (by 6. of this Book) the line MN parallel to AF meets the ſurface (within which the point P is taken) in two points, ſuppoſe in M, N; conſequently in theſe points it meets the parabola by Cor. 1. to the Definitions.

PROP. XVI.

Through any point of a conic ſection one right line only can be drawn, which will touch the ſection in that point.

Fig. 9. LET BAD be a cone, and MVN any conic ſection in its ſurface; let any point V be taken in it, and through V draw the ſide AVB of the cone, and through the point B where the ſide AVB meets the circumference of the baſe draw BK touching the baſe; then the plane ABK touches the ſurface (Cor. 2. 3. of this Book) and the common interſection VX of the plane ABK, with the plane of the ſection, touches the ſection in the point V.

For the line VX is in the plane ABK touching the ſurface, and meets the line of contact in the point V; therefore in this point it touches

touches the furface; but VX is likewife in the plane of the fection MVN which it therefore touches in the fame point V (by Cor. 1. to Definitions). Then if another right line, in the plane of the fection, befides VX, be fuppofed to touch the fection MVN in the point V, it would likewife touch the conical furface in this point: and fince this line is not in the plane ABK, the plane paffing through it and the line AB will not be the fame with the plane ABK, and will touch the furface in the right line AB (by Cor. 3. 3. of this Book) and then two planes would touch the furface in the fame right line (contrary to Cor. 4. 3. of this Book) wherefore one right line only can touch a conic fection in the fame point.

Cor. Hence any right line MN, drawn through a point P within a conic fection, and parallel to any tangent VX, meets the fection in two points.

For let AF be the common interfection of the tangent plane ABK, in which the tangent VX is drawn, with the vertical plane ADE, it will be parallel to the tangent VX (by 16. 11.) and confequently to the line MN drawn through P: but the line AF, becaufe it is in the tangent plane and not the fame with AB the line of contact, falls wholly without the furfaces; therefore the line drawn through P meets the furface (within which the point P is taken) in two points, fuppofe in M, N, (by the 6. of this Book) and therefore meets the fection in the fame points (by Cor. 1. to the Definitions.)

PROP. XVII.

Let MN be a right line terminated by a conic fection; one right line only parallel to this line can touch the fection, if it be a parabola; and if it be an hyperbola, one right line only parallel to MN, can touch it, and likewife one only the oppofite hyperbola; and if the fection be an ellipfe, there are but two right lines parallel to MN which can touch it. Fig. 19. 20, 21.

For let the fection MVN be a parabola, by which the right line MN is terminated; and through A the vertex of the cone, Fig. 19.

cone, let the right line AF be drawn in the vertical plane and parallel to MN, it will (by Cor. 3. 6. of this Book) fall without the conical surfaces, and because the line AF is in the plane ADE touching the surface, through AF another plane ABK may be drawn (by the 4. of this Book) which will touch the surface in a right line as AB: this plane will necessarily cut the plane of the parabola, suppose in the line VX; this line touches the parabola in V (by the preced.) and will be parallel (16. 11.) to the line AF, and consequently to MN terminated by the parabola.

But if (the rest remaining as above) any other right line YZ, parallel to MN, be supposed to touch the parabola; through the vertex A and the point of contact Y draw a side of the cone: the tangent YZ (by hypothesis) is parallel to AF, and (by Cor. 3. 3. of this Book) the plane passing through these parallels touches the surface in the line AY; but this plane will necessarily be different from the tangent planes ADE, ABK; then three planes touching the surface would pass through the line AF contrary to (Cor. 4. of this Book).

FIG. 20, 21. Now let the section MVN be an ellipse or hyperbola, and *mLn* the opposite hyperbola, and let MN be the right line terminated by the ellipse or hyperbola: through the vertex A of the cone let the line AF be drawn in the vertical plane, and parallel to MN, falling (by Cor. 4. 6. of this Book) without the surface, and through AF let two planes ABK, AGH be drawn touching the surfaces in the lines AVB, ALG (by 4. of this Book). Since these planes cut the vertical plane ADE in the line AF, they will likewise cut the plane of the ellipse, or of the opposite sections, in two lines VX, LT parallel to AF (16. 11.) but these lines VX, LT touch the ellipse, or opposite hyperbolas (by the preced.) and because they are parallel to AF they will be parallel to MN.

If then (the rest remaining as before) any third line parallel to MN be supposed to touch the ellipse or hyperbola, in the same manner, as in the former case, it may be shewn that three planes touching the surface might be drawn through AF, which is absurd (Cor. Prop. IV. of this Book).

COR.

Cor. 1. Hence it appears that if any right line touches an ellipse or hyperbola, another right line may be drawn which shall touch the same ellipse, or opposite hyperbola, and be parallel to the other tangent.

Cor. 2. Hence it likewise appears, that if from a point in a parabola a right line be drawn parallel to a tangent, it meets the parabola again; for it cannot touch it (by the first part of this Prop.) therefore it falls on one side within the parabola, and therefore (by Cor. to preced. Prop.) it meets it again. In like manner it may be shewn that a right line drawn from a point in an ellipse, or hyperbola, and parallel to two tangents, meets the ellipse, or hyperbola again.

THE reader is desired to observe that the circle is reckoned among the Conic Sections, for whatsoever is proved universally in what follows of the three sections already defined, obtains also in the circle, as generally appears from the elements of plain Geometry.

PROP. XVIII.

If two right lines meeting each other, be always parallel to two right lines given in position; according as they both touch or cut, or one of them touches and the other cuts a conic section or opposite sections; the squares of the segments of the tangents, or the rectangles under the segments of the secants between the point of concourse of the two lines, and the section, or sections, will be in a constant ratio to each other, wheresoever the point of concourse of the right lines be taken.

THIS Proposition is evident from (Prop. XI. of this Book) and Cor. 1. to the Definitions.

FIG. 22, 23. COR. 1. Hence, if any right line LCV cutting a conic ſection, or oppoſite ſections, in the points L, V, meets two parallel lines in the points O, S cutting the ſame ſection, or oppoſite ſections, or both the oppoſite ſections, in the points A, B and H, G; the rectangles BOA, HSG under the ſegments of the parallel lines between the ſection or ſections, and the line LCV, will be to each other as the rectangles VOL, VSL under the ſegments of the line LCV between the parallel lines and ſection or ſections.

For becauſe BA, LVO are parallel to the ſame lines to which HG, LVO are parallel; in whatſoever ratio the rectangle BOA is to the rectangle VOL, in the ſame ratio the rectangle HSG will be to the rectangle VSL, by this Propoſition: therefore by permutation, BOA is to HSG as VOL to VSL.

COR. 2. Or if any right line cutting a ſection, or oppoſite ſections, meets two parallel lines touching the ſame ſection, or oppoſite ſections, in the ſame manner it may be ſhewn that the ſquares of the ſegments of the parallel tangents between their points of contact and the ſecant which they meet, are to one another as the rectangles under the ſegments of that ſecant between the parallel lines and ſection, or ſections.

COR. 3. Or if any right line cutting a ſection, or oppoſite ſections, meets two parallel right lines, one of them a tangent, and the other a ſecant; the ſquare of the ſegment of the tangent, and the rectangle under the ſegments of the ſecant between the ſection and the ſecant which they meet, will be to each other as the rectangles under the ſegments of that ſecant between the parallel lines and ſection, or ſections.

COR. 4. Or if any right line touching a conic ſection meets two parallel right lines cutting the ſection, or ſections; the rectangles under the ſegments of the ſecants between the ſection, or ſections, and the tangent which they meet, will be to each other as the ſquares of the ſegments of the tangent between the parallel lines, and the point of its contact.

COR. 5. Or if any right line touching a conic ſection meets two parallel right lines touching the ſame ſection, or oppoſite ſections; the ſquares of the ſegments, of the parallel tangents between their

points

points of contact and the tangent which they meet, will be to each other as the squares of the segments of that tangent between the parallels and its point of contact.

Cor. 6. Or if any right line touching a conic section, meets two parallel lines, one of them a tangent and the other a secant; the square of the segment of the tangent, and the rectangle under the segments of the secant between the section and tangent which they meet, will be to each other as the squares of the segments of that tangent between the parallels and its point of contact.

PROP. XIX.

If two right lines parallel to each other meet a right line parallel to any side of a cone, and if both touch or cut, or if one of them touches, and the other cuts a conic section or opposite sections; the squares of the segments of the tangents, or the rectangles under the segments of the secants between the right line, which they meet, and the section or sections, will be to each other as the segments of that line between the parallels, and the point in which it meets the conic section.

FOR since this line, which the parallels meet, is parallel to a side of the cone, it will meet with the surface of the cone, and consequently with the conic section (in the plane of which it is) in one point only (by Prop. V. and Cor. 1. Def.) and therefore this proposition follows from Prop. XII. and Cor. 1. to the Definitions. Fig. 14.

Cor. Hence if MVN be a parabola in the surface of a cone, and the parallel right lines MN and AB cut it, meeting its diameter LV in D, C; the rectangles MDN, ACB under the segments of the parallels between the diameter and parabola, will be to each other as the segments DV, CV of the diameter between the parallels and the vertex V; or if the right line XT parallel to MN touches the parabola in T, and meets the diameter LV in X; the rectangle MDN will be to the square of TX as DV to XV.

For by Definition 15, every diameter of a parabola is parallel to a side of the cone, viz. to that in which the vertical plane touches the surface of which the parabola is a section. Whence the Corollary is evident by this Proposition.

PROP. XX.

Let two right lines parallel to each other, touch an ellipse, or opposite hyperbolas, and in those tangents let two points be taken equidistant from their points of contact. If the right line joining these points cuts the ellipse, or opposite hyperbolas in two points; its segments between these points and the tangents will be equal: or if it touches an ellipse, the segments between its point of contact and the tangents, will be equal.

THIS Proposition is evident from Prop. XIII. and Cor. 1. Def.

PROP. XXI.

Let two right lines, parallel to each other, touch an ellipse, or opposite hyperbolas; a right line joining their points of contact will bisect all right lines which are parallel to the tangents, and terminated by the ellipse, or either of the hyperbolas.

FOR all right lines parallel to the tangents, necessarily meet the line which joins their points of contact; therefore this Proposition follows from Prop. XIV. and Cor. 1. to Def.

Fig. 22, 23. Cor. 1. Hence if two right lines AB, GH parallel to each other, be terminated both ways by an ellipse, hyperbola, or different and opposite hyperbolas; the right line OS bisecting both these, meets the

the ellipse, or hyperbolas in two points; through which points, if tangents be drawn, they will be parallel to the lines bisected.

For let two right lines VX, LT parallel to AB, GH be conceived to touch the ellipse, or opposite hyperbolas (by Prop. XVII.) in the points V, L; then the right line LV joining their points of contact, will bisect the lines AB, GH and consequently will coincide with OS: therefore the Corollary is evident.

Cor. 2. Hence if the right lines AB, GH parallel to each other, and terminated by an ellipse, hyperbola, or different and opposite hyperbolas, be bisected by the right line OS; and if through the points V and L in which the bisecting line meets the ellipse or hyperbolas, two right lines VX, LT be drawn parallel to the bisected lines, they will be tangents; for if not (by preced. Cor.) tangents drawn through these points will be likewise parallel to the lines AB, GH, which is absurd.

Cor. 3. Hence appears a method of drawing two right lines, which shall touch an ellipse or opposite hyperbolas, and be parallel to a right line given in position, cutting the ellipse or hyperbola: for let another secant be drawn parallel to the given secant, and bisect these parallels, and through the points in which the bisecting right line meets the ellipse or hyperbolas, draw two right lines parallel to the line given in position; these lines will be tangents (by Cor. preced.)

PROP. XXII. PROBL. II.

To find the center of an ellipse, or opposite hyperbolas, given in position; and to demonstrate that there can be but one center only of an ellipse or hyperbola.

LET VAGHB be an ellipse or hyperbola, and *m*L*n* its opposite hyperbola, and let two right lines be drawn parallel to each other, cutting the ellipse or hyperbola in the points A, B and G, H, and let the right line OS be drawn bisecting the parallels AB, GH, and meeting (by Cor. 1. preced.) the ellipse or opposite hyperbolas Fig. 22, 23.

 in

in two points, suppose in V, L, and bisect the line VL in C; that point will be the center. For through the points V, L draw two lines VX, LT parallel to AB, GH, they will be tangents (by Cor. 2. preced.) therefore draw through the point C any line meeting the ellipse or opposite hyperbolas in the points E, F and the tangents VX, LT in Q and P; then because the triangles CVQ, CLP are similar, and the lines CV, CL equal; the lines CQ, CP will be equal, and likewise the tangents VQ, LP; therefore QE, PF are equal (by 20. of this Book) and therefore CE, CF are equal. But if the line drawn through C, and meeting the ellipse in two points M, N be parallel to the tangents VX, LT, it will be bisected in C (by the preced. Prop.) therefore since every right line drawn through the point C, and terminated by the ellipse or opposite hyperbolas is bisected in C, this point C will be the center of the ellipse or opposite hyperbolas by Definition XVI. *Q. E. I.*

FIG. 23. If (the above remaining as premised) another point D be supposed to be the center of the hyperbolas, draw through this point a line parallel to VL, it will meet the opposite hyperbolas in two points A, G (Cor. 4. Def.) let it meet the tangents VX, LT in Q, R; VQ, LR will be equal (being opposite sites of a parallelogram) and therefore QA, RG are equal (by Prop. XX. of this Book) but because D is supposed to be the center, DA, DG are equal, therefore DQ, DR are equal: draw any other line through D, meeting the hyperbolas in the points Y, Z, and the tangents in I, K; because D is a center (by hypoth.) DY, DZ are equal, and because the triangles DQI, DRK are similar, and the lines DQ, DR equal, DI, DK will be equal, and consequently (5. 2.) the rectangles YIZ, YKZ are equal: but these rectangles are as the squares of the tangents VI, LK (by Cor. 2. Prop. XVIII.) therefore these squares, and consequently the lines VI, LK are equal; therefore the line IK is parallel to VL (33. 1.) that is to AG; which is absurd; and therefore no point beside C can be the center of the hyperbolas; which may in like manner be demonstrated of the ellipse; but this is self-evident, for if there were two centers in an ellipse, a line terminated by the ellipse and passing through the

the centers, would be bisected in two different points; which is absurd.

Cor. 1. All right lines joining the points of contact of two parallel right lines touching an ellipse or opposite hyperbolas, bisect each other in one and the same point, viz. in the center. For if they could be bisected in different points, there would be several centers of the ellipse or opposite hyperbolas, by the first part of the preceding demonstration; the contrary to what is shewn in the second part.

Cor. 2. Any right line as LV joining the points of contact of two parallel tangents LT, VX passes through the center, and consequently is a diameter of an ellipse, or a transverse diameter of an hyperbola, by Def. XVII.

Cor. 3. Hence two tangents LT, VX, drawn through the two vertices of the same diameter LV of an ellipse or hyperbolas are parallel; for if not, another tangent may be drawn parallel to one of them (Cor. 1. 17. of this Book) and the right line joining the points of contact of these parallels would also pass through the center (Cor. 2. of this Prop.) which is absurd.

Cor. 4. No right line touching an hyperbola passes through the center; for if a right line touches an hyperbola, another right line parallel to it may be drawn to touch the opposite hyperbola (Cor. 1. 17. of this Book) and the right line joining their points of contact passes through the center, and consequently neither of the tangents can pass through the center.

Cor. 5. Hence any right line, drawn from the center of the hyperbola to any point V in it, will be a transverse semidiameter; for it does not touch the hyperbola in V by the preceding Corollary; therefore through the point V let VX be drawn touching the hyperbola, and LT parallel to it touching the opposite hyperbola; it appears that the line joining the points L, V is a transverse diameter (by Cor. 2.) and consequently coincides with the line CV, therefore this line is a transverse semidiameter.

PROP.

PROP. XXIII.

Every right line drawn through the center of an hyperbola, and parallel to a right line touching the hyperbola, or cutting it in two points, is a ſecond diameter.

FIG. 23. THROUGH C the center of the hyperbolas BVA, *m*L*n* let the right line MN be drawn parallel to any tangent VX, and draw through the point of contact the tranſverſe diameter VCL, and through the other vertex L draw LT, touching the oppoſite hyperbola, and parallel to VX (by Cor. 3. preced.) and let any right line be drawn parallel to VCL, and meeting the oppoſite hyperbolas (by Cor. 3. Def.) in two points A, G, the tangents VX, LR in Q, R, and the line MN in D; then the tangents VQ, LR are equal, being oppoſite ſides of a parallelogram, and therefore the lines QA, RG are equal (Prop. XX.) and becauſe VC, CL are equal, and CD parallel to VX, LT; the lines QD, DR will be equal; therefore the line AG will be biſected in D by the line MN: and therefore this right line is a ſecond diameter (by Def. 18.) *Q. E. D.*

PROP. XXIV.

A diameter of any conic ſection biſects all right lines terminated by the ſection which are parallel to the tangent drawn through its vertex.

FIG. 19. FIRST let BAD be a cone, and the ſection MVN in its ſurface a parabola, ADE its vertical plane, and LV a diameter, and let the line VX touch the parabola in the vertex V of the diameter LN, and let MN be parallel to VX, and terminated by the parabola, and meet the diameter LV in P; it will be biſected in this point.

For draw the ſide of the cone AV, and let the plane ABK paſs through AV and the tangent VX; this plane touches the conical ſurface in the ſide AVB (Cor. 3. 3. of this Book) and becauſe AV cuts

cuts the parallels LV, AD, it will be in the same plane with them (7. 11.) therefore the line LV is in the plane of the triangle BAD: draw in this plane through the point P any line meeting AD in any point Q and AB in O; draw through these points in the tangent planes ADE, ABK, the lines QR, OS parallel to AF, the intersection of the tangent plane ABK with the vertical plane ADE, these lines touch the surface in the points Q, O, and because they are parallel to AF they will be parallel to the tangent VX (by 16. 11.) and likewise to MN (9. 11.) but the line MN is bisected in P by the line joining the points of contact Q, O (by Prop. XIV.) and consequently by the diameter LV.

In the second place, let the section be an ellipse or hyperbola, and because the tangents drawn through the two vertices of any diameter are parallel to each other (Cor. 3. 22. of this Book) this diameter will bisect all right lines parallel to those tangents, and terminated by the ellipse or either of the hyperbolas, as appears from Prop. XXI.

PROP. XXV.

A second diameter MN of an hyperbola bisects all right lines terminated by the hyperbolas, and parallel to the transverse diameter LV joining the points of contact of the tangents VX, LT, which are parallel to that second diameter MN: and on the contrary, if a right line AG be bisected by a second diameter MN, it will be parallel to the transverse diameter LV joining the points of contact of the tangents VX, LT, which are parallel to that second diameter.

Part 1. LET the line AG be terminated by the hyperbolas BVA, *m*L*n* and be parallel to the diameter LV joining the points of contact of the tangents VX, LT which are parallel to MN, and let AG meet the diameter MN in D. It will be bisected in this point. FIG. 23.

For

For let AG meet the tangents in the points Q, R; then the tangents VQ, LR will be equal, being the opposite sides of the parallelogram LVQR; therefore the lines AQ, RG are equal (by Prop. XX.) and because VC, CL are equal, and the line CD is parallel to VQ, LR; the lines QD, DR will be equal; therefore AD, DG are equal; and therefore AG is bisected by the second diameter MN.

Part 2. And if (the rest remaining as above) the line AG be bisected by the diameter MN, it will be parallel to the diameter LV joining the points of contact of the tangents VX, LT which are parallel to MN; for because VC, CL are equal, and the lines VQ, CD, LR parallel; QD, DR will be equal; but, by hypothesis, AD, DG are equal; therefore AQ, RG are equal, and therefore the rectangles AQG, ARG are equal; but these rectangles are to each other as the squares of the tangents VQ, LR (by Cor. 2. Prop. XVIII.) therefore these squares and consequently the lines VQ, LR are equal: wherefore (33. 1.) the lines AQRG and LV are parallel. *Q. E. D.*

Corollaries to the two pceceding Prepositions.

Cor. 1. Every right line terminated by a conic section, and bisected by a diameter, is parallel to the tangent passing through the vertex of that diameter (provided when the section is an ellipse, the bisected line be not a diameter) for if it be not parallel to this tangent, let it be parallel (by 17. of this Book) to a tangent drawn through the vertex of another diameter; then by Prop. 24. it would be also bisected by this other diameter, which is absurd.

Cor. 2. All right lines ordinately applied to the same diameter of a conic section are parallel to each other; for when they are applied to a diameter of a parabola, ellipse, or a transverse diameter of an hyperbola, they will all be parallel to the tangent passing through the vertex of that diameter (by preced. Cor.) and when they are applied to a second diameter of the hyperbola, they will be parallel

rallel to one and the ſame tranſverſe diameter, by Caſe 2. of the preceding Propoſition.

Cor. 3. Hence any right line terminated by a conic ſection or by oppoſite hyperbolas, and parallel to the ordinates of any diameter, will be ordinately applied to the ſame diameter.

Cor. 4. Hence alſo, two right lines terminated by a conic ſection or oppoſite ſections, cannot biſect each other unleſs they meet in the center of the ellipſe or of the oppoſite ſections; for if they could, a diameter of the ſection may be drawn biſecting both of them, and conſequently both would be ordinately applied to the ſame diameter, and therefore (by Cor. 2.) parallel to each other, which is abſurd.

Cor. 5. Hence alſo it appears that if two right lines parallel to each other be terminated by a conic ſection or oppoſite ſections, the diameter which biſects one will biſect the other.

Cor. 6. A right line biſecting two parallel lines terminated by a conic ſection or oppoſite ſections is a diameter; for a diameter biſecting one of them will biſect the other.

Cor. 7. Hence, if a ſegment of a conic ſection be given in poſition, the diameters and center of the ſection may be found: for let two parallel right lines be drawn terminated by the ſegment, the right line biſecting theſe will be a diameter; in like manner another diameter may be found, and if it be parallel to the former, the ſection will be a parabola; but if the diameters cut each other the point of interſection will be the center of an ellipſe, if the ſegment be concave towards the center, and if convex, the ſection will be an hyperbola.

Cor. 8. A right line drawn through the vertex of a diameter parallel to its ordinates, touches the ſection; for if not, let a tangent be drawn through its vertex, this likewise will be parallel to the ordinates of that diameter (by Cor. 1.) which is abſurd.

Cor. 9. Hence appears a method of drawing a right line which ſhall touch a conic ſection given in poſition, and be parallel to any right line cutting the ſection in two points; for let a line be drawn parallel to this ſecant and terminated by the ſection, the right line biſecting theſe parallels will be a diameter, and the right line drawn through the vertex of this diameter and parallel to the biſected lines touches the ſection by the preced. Cor.

PROP. XXVI.

If two right lines touching a conic ſection or oppoſite ſections meet each other; a right line paſſing through their point of concourſe and biſecting the right line joining their points of contact, will be a diameter of the ſection.

Fig. 25, 26. FOR let the right lines PM, PN touch the conic ſection or oppoſite ſections in M, N, and join the points M, N, and draw PO biſecting MN in O; PO will be a diameter: for if not, through the point O let a diameter of the ſection be drawn meeting the tangent PM in Q, and let NQ be drawn meeting the ſection again in A; through the point A draw the right line AD parallel to MN, this meets the oppoſite ſection, ſuppoſe in B (Cor. 4. Def.) when the points M, N are in oppoſite ſections; and when theſe points are in the ſame ſection, becauſe A is not the vertex of the diameter OQ, the line AD parallel to MN does not touch the ſection (by Cor. 1. preced. and Cor. 2. 17. of this Book) let it then meet the ſection again in B and OQ in X and PM in D; ſince the line AB is parallel to MN, which is biſected by the diameter OQ, it will be biſected by the ſame diameter (by Cor. 5. preced.) therefore the ſegment AX will be equal to BX; and becauſe in the triangle MQN the line OQ drawn from the vertex, biſects the baſe in O, the ſame line will biſect AD, parallel to the baſe in X; there-

therefore the ſegment DX is equal to the ſegment AX, that is to BX, which is impoſſible; and therefore no line drawn through O beſide PO will be a diameter of the ſection. *Q. E. D.*

Cor. 1. On the contrary, if two right lines touching a conic ſection or oppoſite ſections meet each other; the diameter biſecting the right line joining the points of contact, will paſs through the point where the tangents meet; for one diameter only can biſect the line joining the points of contact, and a right line biſecting this line joining the contacts and paſſing through the concourſe of the tangents is a diameter; therefore the Corollary is manifeſt.

Cor. 2. Hence it appears that if two right lines touching a conic ſection or oppoſite ſections meet each other, their point of concourſe will be in the diameter which biſects the line that joins their points of contact.

Cor. 3. Hence, three right lines touching a ſection or oppoſite ſections cannot meet in one point. For if it were poſſible, draw two right lines joining one of the points of contact with the other two; then (by Cor. 1.) two diameters biſecting theſe lines would meet each other in the point of concourſe of the tangents, which is impoſſible; for if the ſection be a parabola, the diameters are parallel, and the diameters of an ellipſe or hyperbola meet only in the center, through which no tangent paſſes (Cor. 4. 22. of this Book.)

PROP. XXVII.

Two diameters of an ellipſe or hyperbola, of which each is parallel to the ordinates of the other, are conjugate diameters.

LET MN be a diameter of an ellipſe or a ſecond diameter of an hyperbola, to which let the right lines AG be ordinately applied meeting it in the point D, and let LV be the diameter pa- Fig. 22, 23.

 rallel

rallel to AG; these two diameters MN, LV will be conjugate diameters.

For let the tangents VX, LT be drawn through the two vertices of the diameter LV; they will be parallel (by Cor. 3. Prop. XXII.) and let them meet AG (if necessary produced) in the points Q, R; then the tangents VQ, LR will be equal, being opposite sides of a parallelogram, and therefore (by Prop. XX.) the right lines AQ, RG will be equal; but AD, DG (by hypothesis) are equal; therefore the lines QD, DR are equal, and likewise VC, CL are equal; therefore CD, or MN is parallel to VQ, LT; but the diameter LV bisects (by 24. of this Book) all right lines which are parallel to VQ, LT, that is to the diameter MN; therefore the diameters MN, LV are conjugate diameters (by Def. XIX.)

Now let the right line AB be ordinately applied to the transverse diameter LV of an hyperbola, and draw the second diameter MN parallel to AB; LV, MN will be conjugate diameters. For through the two vertices of the diameter LV let the lines VX, LT be drawn parallel to AB, that is to the diameter MN; they will be tangents (Cor. 8. 25. of this Book) therefore by Prop. XXV. the diameter MN bisects all right lines terminated by the opposite hyperbolas and parallel to the diameters LV; and therefore LV, MN are conjugate diameters. *Q. E. D.*

Cor. It appears from hence that an ordinate to any diameter is parallel to its conjugate diameter, and that two diameters are conjugate, if one of them be parallel to the tangent drawn through the vertex of the other.

And on the contrary a tangent drawn through the vertex of a diameter is parallel to its conjugate.

PROP.

PROP. XXVIII. PROB. III.

A conic ſection and its diameter being given in poſition; through a given point in the ſection to draw a right line which ſhall be ordinately applied to that diameter.

FIRST let the ſection be a parabola or an hyperbola and MN the diameter given in poſition (which in the hyperbola muſt be a tranſverſe diameter) and let A be the point given in the ſection, to the diameter MN let any right line AN be drawn and produced to H, ſo that AN, NH be equal; through H let a right line be drawn parallel to MN which (by Cor. 4, 5. to Def.) will meet the ſection in ſome point as G; the right line joining the points A, G will be ordinately applied to the diameter MN. FIG. 27.

For becauſe AN, NH are equal and DN, GH parallel; AD, DG will be equal.

Secondly, let the ſection be an ellipſe or hyperbola, and MN any diameter of the ellipſe or a conjugate diameter of the hyperbola, to which, from the point A in the ſection, an ordinate is to be applied: find the center of the ſection (22. of this Book) and draw through the point A the diameter ACH; if the right line drawn through its vertex H or A parallel to MN touches the ſection, then the diameter ACH will be conjugate to MN, and theſe diameters are ſaid to be ordinately applied to each other: but if the line drawn through H and parallel to MN meets the ſection again in any point G; the line joining the points A, G will be an ordinate to the diameter MN; for let it meet it in D; becauſe AC, CH are equal and HG, CD parallel; AD, DG will be equal; therefore the line AG is ordinately applied to the diameter MN. *Q. E. F.* FIG. 22, 23.

COR. Hence, a line may be drawn through a given point in a conic ſection given in poſition to touch the ſection: for (by Cor. 7. Prop. XXV.) let two diameters of the ſection be found, and through the given point draw a diameter to which an ordinate may be applied (by this Prop.) the line parallel to this ordinate drawn through the given point will touch the ſection (by Cor. 8. Prop. XXV.)

PROP.

PROP. XXIX.

The squares of right lines which are ordinately applied to the same diameter of a parabola, are as the abscisses of that diameter between the ordinates and vertex.

FIG. 28. LET LV be a diameter of a parabola and L its vertex, and let it meet the lines BA, HG ordinately applied to it in the points C, D.

Because these lines are parallel to each other (by Cor. 2. Prop. XXV.) the rectangles BCA, HDG, that is the squares of the ordinates BC, HD are (by Cor. Prop. XIX.) to each other as CL, DL, viz. the abscisses of the diameter between the ordinates and the vertex L (Def. XXI.) *Q. E. D.*

PROP. XXX.

The squares of right lines, ordinately applied to any diameter of an ellipse or to a transverse diameter of an hyperbola, are to each other as the rectangles under the abscisses of the diameter between the two vertices and ordinates.

FIG. 22, 23. LET LV be a diameter of an ellipse or a transverse diameter of an hyperbola and the points L, V its two vertices, and let it meet in the points O, S two ordinates BA, HG terminated by the ellipse, hyperbola or opposite hyperbolas.

Because these right lines are parallel (Cor. 2. Prop. XXV.) the rectangles BOA, HSG (by Cor. 1. Prop. XVIII.) that is the squares of the ordinates BO, HS, will be to each other as the rectangles LOV, LSV under the abscisses of the diameter LV between the ordinates and the two vertices. *Q. E. D.*

COR. 1. As it is manifest by this Proposition, that in the ellipse or opposite hyperbolas the ordinates applied to the same diameter and at equal distances from the center, are equal; it is likewise evident, that if the opposite sections be so placed that the line VO may coincide with LS, and the angle VOB with its alternate angle LSG, the whole

whole ſection will be congruous: and alſo the two ſegments of an ellipſe into which it is divided by its diameter, may be ſo placed as wholly to coincide.

Cor. 2. Hence, if a circle be deſcribed upon a diameter of an ellipſe and from two points in this diameter ordinates applied to this diameter be drawn to the ellipſe, and at the ſame points two perpendiculars be erected meeting the circumference of the circle: the ordinates will be to each other as theſe perpendiculars: for by this Prop. the ſquares of the ordinates are to each other as the rectangles under the abſciſſes of the diameter between theſe ordinates and its two vertices, that is (from a property of the circle) as the ſquares of theſe perpendiculars; therefore theſe ordinates are as the perpendiculars.

PROP. XXXI.

If two right lines meeting each other touch or cut, or one of them touches and the other cuts an ellipſe; the ſquares of the ſegments of the tangents, or the rectangles under the ſegments of the ſecants between the point where the right lines meet and the ellipſe, will be to each other as the ſquares of the ſemidiameters to which they are parallel.

FOR the diameters of an ellipſe are ſecants which meet each other in the center, and the rectangles under their ſegments between the center and the ellipſe are the ſquares of the ſemidiameters; therefore (by 18. of this Book) the ſquares of the tangents, or the rectangles under the ſegments of the ſecants, are to each other as the ſquares of the ſemidiameters to which they are parallel.

Cor. 1. Hence, if in an ellipſe a right line AD be ordinately applied to the diameter MN; the rectangle MDN under the abſciſſes of the diameter MN will be to the ſquare of the ordinate AD as the ſquare of the diameter MN to the ſquare of its conjugate LV: for let AD produced meet the ellipſe in G, the ſquare of AD is Fig. 22.

equal

equal to the rectangle ADG, and (by Cor. 27. of this Book) the diameter LV and right line AG are parallel; wherefore the Corollary appears from this Proposition.

Cor. 2. If two conjugate diameters MN, LV of an ellipse are perpendicular to each other, they will be unequal; for if they were equal, from any point A of the ellipse, let the right line AD be ordinately applied to the diameter MN, it will be parallel to LV (Cor. 27. of this Book) and from hence, MDA is a right angle (by hypothesis) let C be the center and join C, A; the square of CA (by 47. 1.) is equal to the squares of CD, DA, that is to the square of CD, together with the rectangle MDN (for the rectangle MDN and the square of DA are equal, by the preced. Cor. because the diameters MN, LV are equal by hypothesis) or to the square of the semidiameter CN (5. 2.); therefore the semidiameters CA, CN are equal; and in like manner it may be shewn that all the other semidiameters would be equal to each other and then the ellipse would be a circle, which is absurd; and therefore this Corollary is manifest.

Cor. 3. Hence, it appears, that if two right lines touching an ellipse meet each other; the segments of the tangents between their point of concourse and the points of contact, will be to each other as the semidiameters to which they are parallel.

DEFINITION XXIV.

Fig. 23. IF in a second diameter of an hyperbola, a segment MN bisected in the center be taken so that the square of the segment MN, be to the square of the diameter LV conjugate to this second diameter, as the squares of the ordinates applied to the diameter LV, to the rectangles under the abscisses; the points M, N the extremities of this segment are called the *two vertices* of that second diameter: and the magnitude of any second diameter is determined by these points, as the magnitude of a transverse diameter is determined by its two vertices.

PROP.

PROP. XXXII.

If from a point in an hyperbola a right line be ordinately applied to a second diameter; the square of this second diameter is to the square of the transverse diameter conjugate to it, as the sum of the squares of the second semidiameter and of its segment, between the center and ordinate, will be to the square of the ordinate.

FROM the point A in an hyperbola, let AD be ordinately applied to the second diameter MN, and let the diameter LV be conjugate to MN; the square of CN is to the square of CV, as the squares of CN, CD together to the square of AD. *Fig.* 23.

For let AO be an ordinate to LV, COAD will be a parallelogram (by Cor. Prop. XXVII.) and consequently AD, OC and likewise AO, DC are equal.

By the preceding Definition, the square of CN is to the square of CV as the square of AO or CD to the rectangle VOL; therefore (by 12. 5.) the square of CN is to the square of CV as the sum of the squares of CN, CD to the square of CV and the rectangle VOL together, that is (6. 2.) to the square of CO or AD. *Q. E. D.*

Cor. Hence, if from two points in an hyperbola or opposite hyperbolas, two right lines be drawn ordinately applied to a second diameter; the square of the first line drawn will be to the square of the other as the sum of the squares of the second semidiameter and of the distance of the first line from the center to the sum of the squares of the same second semidiameter and of the distance of the other line from the center.

PROP. XXXIII.

Fig. 29. Any indefinite right line VL being given in position, and a point V given in it; and likewise the right line MN given, bisected by VL in the point P; a parabola may be described, of which the line VL shall be a diameter, V its vertex, and to which MN shall be ordinately applied.

FOR let any plane different from the plane MPV pass through the line MN, and in this plane describe a circle about the diameter MN and draw the diameter BPD perpendicular to MN, meeting the circle in the points B, D; let B, V be joined, and through the point D in the plane VBD draw the right line DA parallel to VL which may meet somewhere BV, suppose in the point A; whilst this point remains fixt, let the line AD be moved round the periphery of the circle BNDM; the intersection of the conical surface, generated by this motion, with the plane MPV, will be a parabola.

For through the point D let DE be drawn parallel to MN, it will be perpendicular to the diameter BD (by construction) and therefore touches the circle in D, and hence the plane ADE touches the surface (3. of this Book) and will be parallel to the plane of the section MVN (15. 11.) this section is therefore a parabola and the line VL parallel to AD a diameter whose vertex is the point V, and the line MN an ordinate applied to this diameter; all which are evident from the 11th and 15th Definitions. *Q. E. D.*

Cor. If two parallel right lines NP, YZ meet any right line VL, and a point V be taken in it (but not between P and Z) so that the square of NP be to the square of YZ as PV to ZV; the points N, Y will be in a parabola to which LV is a diameter, and V its vertex, and to which NP, YZ, are ordinately applied; for if (by this Prop.) a parabola be described about the diameter VL whose vertex is V, and the line NPM be ordinately applied to that diameter; then because the line YZ cuts the diameter below the vertex,

TAB. 3.
Page 42.

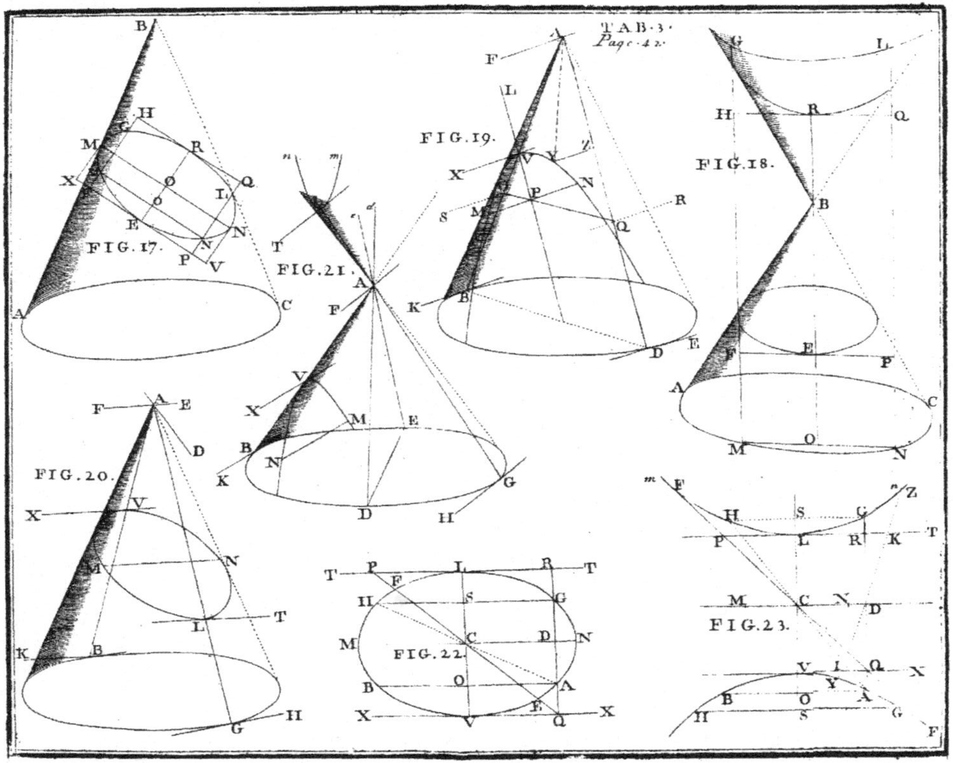

FIG. 17. FIG. 18. FIG. 19. FIG. 20. FIG. 21. FIG. 22. FIG. 23.

tex, produced both ways, it will cut the parabola (15. of this Book) in two points, and will be ordinately applied to the diameter LV (Cor. 3. Prop. XXV.) and therefore the point Y will be in this parabola; for if the parabola meets the line ZY produced, or between the points Z and Y (by Prop. XXIX.) the square of NP would be to the square of a line either greater or less than YZ, as PV to ZV, which is contrary to the hypothesis.

LEMMA I.

IF in two planes intersecting each other two right lines be parallel; they will be parallel to the common intersection of the planes.

For, if not, they will meet the intersection of the planes somewhere, and then the intersection of the planes would be in the same plane with the two parallel right lines (by 7. 11.) contrary to the hypothesis.

PROP. XXXIV.

Two right lines bisecting each other being given; an ellipse may be described in which the given lines shall be conjugate diameters; but the lines must not be equal, if they be at right angles to each other.

LET VL, MN be the two given lines, and C the point of intersection, and let any plane different from the plane MCV pass through the line MN, and through the point C, in this plane draw the right line BD perpendicular to MN, and in this line let any other point P, beside C, be taken as a center, and a circle described passing through the points M, N, meeting BD somewhere as in the points B, D. Let the lines joining the points B, V and L, D meet each other in A (for they are in the same plane with the lines BD, VL (2. 11.) and will not be parallel, for if they were, the triangles VCB, LCD would be similar, and therefore because CV, CL are equal, BC, CD would be equal contrary to the hypothesis.) Let the line AD, whilst the point A remains fixt, be carried round the periphery of the circle BNDM: the intersection of the conical surface, generated by this motion, with the plane MVNL, FIG. 30.

 will

will be an ellipse, of which the lines VL, MN are conjugate diameters.

Through the point D, the extremity of the diameter BD draw DE a tangent to the circle, it will be parallel to MN (by construction, and 16. 3.) and the plane ADE touches the surface in the line ADL, and the line LT the intersection of this tangent plane with the plane of the section VNLM, touches the section in L; and since the lines MN, DE in these intersecting planes are parallel to each other, they will (by Lemma 1.) be parallel to the common intersection LT; but since the lines VL, MN are terminated by the section MVNL and bisect each other in the point C; they will be diameters of this section, whether it be an ellipse (Cor. 4. 25. of this Book) or a circle. If the diameters VL, MN be not perpendicular to each other, the tangent LT parallel to MN will not be perpendicular to the diameter LV drawn through the point of contact; therefore the section will not be a circle (16. 3.) and if the diameters VL, MN be perpendicular to each other, by hypothesis, they are unequal; therefore in this case the section MVNL will not be a circle, and therefore the section surrounding the cone is an ellipse (by Cor. 7. to the Defin.) and the diameters VL, MN are conjugate (Cor. 27. of this Book) because MN is parallel to the tangent LT drawn through the vertex of the diameter VL. *Q. E. D.*

Cor. Let the given right lines VL, MN bisect each other in C, as in the proposition: if YZ be drawn from any point Y to the line VL, and parallel to MN, and if the rectangle LZV be to the square of YZ as the square of LC to the square of MC; the point Y will be in the ellipse described about the conjugate diameters VL, MN. For if not, the rectangle LZV would be to the square of a line greater or less than YZ, as the square of LC to the square of MC (Cor. 1. Prop. XXXI.) contrary to the hypothesis.

PROP.

PROP. XXXV.

Opposite hyperbolas may be described about two given right lines bisecting each other, one of which lines shall be a transverse diameter, and the other a second diameter conjugate to it.

LET VL, MN be the given lines intersecting each other in C; let any point O be taken in the line LV produced, and through this point draw RQ parallel to MN, and take on each side of it RO equal to OQ, so that the square of RO or OQ may be to the rectangle VOL, as the square of MN to the square of LV; let any plane, different from the plane ROV, pass through the line RQ, and in this plane draw through the point O the right line BD perpendicular to RQ, and take in this line any point P for a center, and describe a circle passing through the points R, Q which may meet the line BD somewhere as in the points B, D, and join BV and DL: these lines BV, DL meet somewhere, suppose in A (for they are in the same plane with BD, VL and are not parallel) let a line drawn through the point A and produced both ways, revolve about the periphery of the circle BQDR; the intersections of the opposite surfaces, which are generated by the motion of this line, with the plane MVNL, will be opposite hyperbolas by Def. XIII. FIG. 31.

Through the points B, D, let BK, DE be drawn touching the circle, they will be parallel to RQ which is perpendicular to the diameter BD; then VX the intersection of the tangent plane ABK with the plane of the hyperbolas touches the hyperbola RVQ in V, and (by Lemma 1.) it will be parallel to OR, BK; and LT the intersection of the tangent plane ADE with the plane of the hyperbolas, touches the opposite hyperbola in L, and will be parallel to OR, DE (by Lemma 1.) therefore the tangents VX, LT are parallel to each other and to the line MN, and therefore the line VL is a transverse diameter of the hyperbolas (by Cor. 2. Prop. XXII.) and

and MN a ſecond diameter conjugate to VL (Cor. Prop. XXVII.) and the points M, N, are its two vertices, by Conſtruction and Definition XXIV. In the ſame manner, two other oppoſite hyperbolas may be deſcribed, of which MN ſhall be a tranſverſe diameter and VL a ſecond diameter conjugate to it. *Q. E. D.*

Cor. Let the given lines VL, MN biſect each other, and to the line VL produced draw RO parallel to MN, ſo that the ſquare of MN will be to the ſquare of VL as the ſquare of RO to the rectangle VOL; the point R will be in the hyperbola of which VL is a tranſverſe diameter to which RO is ordinately applied, and MN will be its conjugate, as is evident.

DEFINITION XXV.

Fig. 32. LET BAF be a cone, and the ſections *a*V*t*, *m*L*n* oppoſite hyperbolas, and ADE the vertical plane meeting the oppoſite ſurfaces in the right lines DA*d*, EA*e*: draw the planes ADGC, AEHC touching the conical ſurfaces in the right lines DA*d*, EA*e*, and let GC*g*, HC*h* be the interſections of theſe planes with the plane of the hyperbolas; theſe lines are called the *Aſymptotes* of the hyperbolas.

Cor. 1. The aſymptotes GC*g*, HC*h* of the hyperbolas are parallel to the ſides of the cone AD, AE, in which lines the vertical plane meets the ſurface; for the vertical plane is parallel to the plane of the hyperbolas; therefore the Corollary is evident from this Definition and (16. 11.)

Cor. 2. The aſymptotes GC*g*, HC*h* do not meet the hyperbolas; for the aſymptotes are in the planes touching the conical ſurface, and (by Cor. preced.) are parallel to the lines DA*d*, EA*e* in which lines only theſe planes meet the ſurface; therefore the aſymptotes do not meet the conical ſurfaces and therefore do not meet the hyperbolas which are in thoſe ſurfaces.

Cor.

Cor. 3. Every right line as NV drawn in the plane of the hyperbolas parallel to either of the asymptotes HC*b* meets one of the hyperbolas in one point only, and will be wholly on one side within this hyperbola, and on the other part wholly without both of them. For (by Cor. 1.) the line NV is parallel to the side AE of the cone; therefore (by Prop. V.) it meets one of the opposite surfaces in one point only, and on one part falls wholly within this surface, and on the other part wholly without both surfaces; therefore because the line NV is in the plane of the hyperbolas, this Corollary appears from Cor. 1. to the Definitions of the Sections.

Cor. 4. If a right line PQ cuts both the sides which contain the angle GC*b* adjacent to the angle GCH, containing the hyperbola, it will meet both the hyperbolas. For if CV be drawn from the point of intersection C of the asymptotes parallel to PQ, it falls within the angle GCH, and therefore because the angles GCH, DAE are equal (10. 11.) if a line be drawn from the vertex A of the cone parallel to CV or PQ, it falls within the angle DAE, and consequently within the conical surfaces; therefore the right line PQ meets the opposite surfaces (by Cor. 2. Prop. VI.) and therefore meets both the hyperbolas, because it is in the same plane with them.

Cor. 5. If a right line touches an hyperbola in any point V, it meets both the asymptotes HC, GC which contain that hyperbola; for if it were parallel to either of the asymptotes it would be on one side of the point V within the hyperbola (by Cor. 3.) contrary to the hypothesis; or if it could meet the sides containing the angle adjacent to HCG, it would meet both the hyperbolas (by Cor. 4.) and therefore would not be a tangent (by Cor. 2. to the Def. of the Sections) therefore it meets the sides of the angles HCG containing the hyperbola.

Cor. 6. It is evident that a right line RS cutting the hyperbola or opposite hyperbolas meets the asymptotes.

PROP.

PROP. XXXVI.

If a right line cutting an hyperbola or oppofite hyperbolas meets the afymptotes in two points; the fegments between the hyperbola or hyperbolas and afymptotes will be equal: Or if a right line touches an hyperbola; the fegments between the point of contact and the afymptotes, will be equal: Or if a right line meets the afymptotes and the fegments between the afymptotes and a point in the hyperbola be equal; it will touch the hyperbola in that point.

Fig. 32. *Part* 1. LET the line RS cut the hyperbola or oppofite hyperbolas in the points R, S and meet the afymptotes in P, Q; the fegments PR, QS will be equal.

For let AC be the common interfection of the tangent planes ADGC, AEHC, and draw through the points P, Q, in thefe planes, the lines PK, QI parallel to AC, meeting the fides of the cone *d*AD, *e*AE in the points K, I, it is manifeft that thefe lines touch the fame or oppofite conical furfaces in the points K, L, and being oppofite fides of parallelograms are equal to AC, and confequently to each other; and therefore fince the right line joining the points P, Q cuts the fame or oppofite furfaces in R, S; the fegments RP, SQ will be equal (by Prop. XIII.)

Part 2. Let any right line as TX touch an hyperbola in V and meet the afymptotes in the points X and T; the fegments VX, VT will be equal; for from the points X, T, draw in the tangent planes two lines XY, TZ, parallel to AC, meeting the fides of the cone *d*AD, *e*AE in the points Y, Z. It is evident that thefe lines touch the conical furface BAF in thofe points and are equal to AC, being oppofite fides of parallelograms, and confequently are equal to each other; therefore fince the line joining the points X, T touches the conical furface in V, the fegments VX, VT will be equal by Prop. XIII.

Part

Part 3. Or if the right line XT meeting the aſymptotes in X, T meets the hyperbola in V, and the ſegments VX, VT be equal; this line touches the hyperbola in the point V.

For if it could meet the ſection in any other point, ſuppoſe in *p*, the ſegment T*p* would be equal to the ſegment VX, by the firſt part, that is to VT, which is abſurd: therefore it does not meet the hyperbola but in the point V and therefore it touches it in this point; for it does not fall on one part wholly within the hyperbola, becauſe by hypotheſis it meets both the aſymptotes. *Q. E. D.*

Cor. 1. The aſymptotes meet in the center of the hyperbolas: for draw two parallel right lines QP, HG cutting the hyperbola in the points S, R and *a*, *t* and meeting the aſymptotes in the points Q, P and H, G, and through C the interſection of the aſymptotes draw CV biſecting the line QP in O; this will likewiſe biſect the line HG (4. 6.) in the point *q*; but as the ſegments RP, SQ, and likewiſe *t*G, *a*H are equal, the line CV will biſect the lines SR, *a t*, terminated by the hyperbola; therefore this line is a tranſverſe diameter of the hyperbola: and in like manner it may be ſhewn that any other line drawn through the interſection of the aſymptotes is a tranſverſe diameter; and therefore the aſymptotes meet in the center of the hyperbolas.

Cor. 2. Every right line LCV drawn through the center and within the angle HCG containing the hyperbola is a tranſverſe diameter. For any line drawn parallel to LCV, and in the plane of the hyperbolas, neceſſarily meets the ſides containing the angle adjacent to HCG, and therefore meets the oppoſite hyperbolas (Cor. 4. to Def. XXV.) therefore LCV parallel to this line meets (Cor. 3. to Def. of the Sections) the oppoſite hyperbolas; but it paſſes through the center and therefore is a tranſverſe diameter by Def. XVII.

Cor. 3. Every right line drawn through the center as MN, and within the angle adjacent to the angle HCG containing the hyperbola is a ſecond diameter. For through any point O within the hyperbola draw a line parallel to MN; it is evident that it meets the ſides containing the angle HCG, and conſequently the hyperbola in two points; therefore the line MN is a ſecond diameter, by Prop. XXIII.

COR. 4. Hence, the aſymptotes being given in poſition, a line may be drawn which ſhall touch an hyperbola in a given point; for let the aſymptotes HC, GC be given in poſition, and the point V given in the hyperbola; through V let the right line V*r* be drawn parallel to HC, one of the aſymptotes, and meeting the other CG in the point *r*, and let *r*X be taken in the aſymptote CG equal to C*r*, and join VX, this line will be a tangent; for let it, produced, meet HC in T; then becauſe CT, *r*V are parallel and C*r*, *r*X equal, TV, VX will be equal; therefore the line TX touches the hyperbola in V, by Caſe 3. of this Prop.

COR. 5. Hence, oppoſite hyperbolas and their aſymptotes being given in poſition; a line may be drawn from a given point X in the aſymptote CG, which ſhall touch the adjacent hyperbola: for let CX be biſected in *r*, and through the point *r* draw a right line parallel to the other aſymptote CH, meeting the adjacent hyperbola in V, the right line joining the points X, V will be a tangent; for let it, produced, meet the aſymptote CH in T, then becauſe X*r*, *r*C are equal, and V*r*, TC are parallel, XV, VT will be equal (2. 6.) therefore the line XVT is a tangent

PROP. XXXVII.

If a right line cutting an hyperbola meets an aſymptote; the rectangle under the ſegments of the ſecant between the aſymptote and hyperbola will be equal to the ſquare of the ſegment of the tangent parallel to this ſecant, intercepted between the point of contact, and the aſymptote.

FIG. 32. LET a right line cut the hyperbola in the points S, R and meet the aſymptote CG in P, and let the right line XT parallel to SR touch the hyperbola in V, and meet the aſymptote CG in X; the rectangle RPS will be equal to the ſquare of VX.

Through

Through the points P, X let two parallel right lines be drawn in the tangent plane ADGC, and meet the side *d*AD of the cone in the points K, Y; they will touch the conical surface in those points and will be equal, because XP, YK are parallel (Cor. 1. Def. XXV.) then because the line PRS cuts and PK touches the conical surface, and these lines are parallel to the two lines XV, XY touching the same surface; the rectangle RPS (by Prop. XI.) will be to the square of PK as the square of XV to the square of XY: but the squares of PK, XY are equal; therefore the rectangle RPS is equal to the square of VX (14. 5.) *Q. E. D.*

Cor. Hence, if through a point R in an hyperbola a right line be drawn meeting the asymptotes in the points P, Q and the same hyperbola again in S; the rectangle PRQ will be equal to the square of the segment of the tangent VX, parallel to PQ, intercepted between the point of contact and asymptote. For because RP, SQ are equal, PS, RQ will be equal; therefore the rectangle PRQ will be equal to the rectangle RPS, that is to the square of XV, by this Prop.

PROP. XXXVIII.

Any right line terminated by the asymptotes and touching an hyperbola, is equal to the conjugate diameter which is parallel to it.

LET a line touch an hyperbola in the point V, and meet its asymptotes in the points T, X; the line TX will be equal to the conjugate diameter MN parallel to TX. Fig. 32.

For through the point of contact draw the transverse diameter LVO, and through a point O taken in it, draw the right line SR ordinately applied to this diameter; this line will be parallel to the tangent VX; let SR be produced both ways that it may meet the asymptotes in the points Q, P; the segments OQ, OP will be equal: and because (by preced. Cor.) the rectangle PRQ is equal

 to

to the square of VX, the square of OR is the difference of the squares of OP and VX (5. 2.) and the rectangle VOL (6. 2.) is the difference of the squares of CO, CV; then because the triangles CVX, COP are similar, the square of OP will be to the square of VX as the square of CO to the square of CV; therefore, by division, the square of OR will be to the square of VX as the rectangle VOL to the square of CV, and therefore, by permutation, the square of OR is to the rectangle VOL as the square of VX to the square of CV, that is as the square of TX to the square of LV: therefore TX is equal to the second diameter conjugate to LV (by Def. XXIV.) and consequently is parallel to TX.

Cor. 1. The right line VN joining the vertices of the conjugate diameters LV, MN, is parallel to one of the asymptotes and bisected by the other. Let the asymptote CG meet VN in *r*, and let CH be the other asymptote; through the vertex V of the diameter VL draw a right line touching the hyperbola and meeting the asymptotes in the points T, X: (by Cor. 27. of this Book) VT will be parallel to CN and will be equal to it (by this Prop.) therefore (33. 1.) the lines VN, TC are parallel and equal; but because XV is half of XT, V*r* will be (2. 6.) half of TC or VN; therefore VN is bisected in the point *r*, and from hence the Corollary is manifest.

Cor. 2. Hence an hyperbola being given in position, the asymptotes are easily found; for let any transverse diameter LV and the center C be found (by Cor. 7. Prop. XXV.) and (by Prop. XXVIII.) draw RO ordinately applied to the diameter LV, and (by Def. XXIV.) let the second diameter MN conjugate to LV be found, and join the vertices V, N of the conjugate diameters, and from the center draw the right lines CG bisecting VN, and CH parallel to VN; the lines CG, CH will be the asymptotes, as is evident.

PROP.

PROP. XXXIX.

If through a point a right line be drawn taken in an aſymptote cutting an hyperbola or oppoſite hyperbolas; the rectangle under the ſegments of the ſecant between the aſymptote and hyperbola or oppoſite hyperbolas, will be equal to the ſquare of the ſemidiameter parallel to that ſecant.

FIRST draw from the point P in the aſymptote CG a right line cutting the hyperbola in the points R, S, and let CN be the ſecond ſemidiameter parallel to PRS; the rectangle RPS will be equal to the ſquare of the ſegment VX of the tangent (by 37. of this Book) that is to the ſquare of CN by the preced. Prop. Fig. 32.

Secondly: draw a right line through the ſame point P cutting the oppoſite hyperbolas in the points R, S, and let LCV be the tranſverſe diameter parallel to RPS, and draw in the tangent plane ADG the line PK, parallel to the interſection CA of the tangent planes ADG, AEH and meeting the ſide AD of the cone in K; it is manifeſt that PK touches the conical ſurface BAF in K, and is equal to CA; then becauſe the line PK touches the conical ſurface and the line PRS cuts both ſurfaces and theſe lines are parallel to the two lines CA, LV meeting in the point C and one of them CA paſſes through the vertex of the cone and conſequently touches the ſurface in A, and the other cuts both the ſurfaces; the rectangle (by XI. of this Book) RPS will be to the ſquare of PK as the rectangle VCL to the ſquare of CA: but the ſquares of PK, and CA are equal; therefore the rectangle RPS (14. 5.) is equal to the rectangle VCL, that is to the ſquare of CV or CL the tranſverſe ſemidiameter parallel to RPS. *Q. E. D.*

Cor. 1. Or if through the point R in an hyperbola a right line be drawn meeting the aſymptotes in two points P, Q; the rectangle PRQ under the ſegments between the hyperbola and aſymptotes,

totes, is equal to the ſquare of the ſemidiameter which is parallel to PQ: for let the line PQ be parallel to the ſecond ſemidiameter CN; then the rectangle PRQ will be equal to the rectangle RPS (by Cor. 37. of this Book) that is to the ſquare of CN, by the firſt caſe of this Prop.

Now let the line PQ be parallel to the tranſverſe ſemidiameter CV, and being produced meet the oppoſite hyperbola in S; then becauſe the ſegments PR, QS are equal, the rectangle PRQ is equal to the rectangle RPS, that is to the ſquare of CV, by the ſecond caſe of this Prop.

Cor. 2. Hence, if a right line as RS terminated by the oppoſite hyperbolas be cut in the points P, Q, ſo that each of the rectangles RPS, RQS be equal to the ſquare of the ſemidiameter CV which is parallel to RS; the points P, Q will be in the aſymptotes, as is evident.

Cor. 3. Hence, likewiſe, a tranſverſe diameter is leſs than any line parallel to it, and terminated by the oppoſite hyperbolas.

Cor. 4. Since any right line touching an hyperbola SVR meets both its aſymptotes CG, CH, any two lines touching the ſame hyperbola, meet each other within the angle GCH; and if two lines not parallel to each other touch the oppoſite hyperbolas, they neceſſarily meet within the angle adjacent to the angle GCH. For two tangents cannot meet each other in an aſymptote; for if they could, the diameter, viz. the line drawn from the point of concourſe of the tangents, and biſecting the line joining their points of contact (26. of this Book) would meet the aſymptote in another point beſide the center, which is abſurd.

PROP.

PROP. XL.

If two right lines meet each other, and according as both touch or cut, or one of them touches and the other cuts an hyperbola or oppofite hyperbolas; the fquares of the fegments of the tangents, or the rectangles contained by the fegments of the fecants between the point of concourfe of the right lines and hyperbola or oppofite hyperbolas, will be to each other as the fquares of the femidiameters to which the right lines are parallel.

FOR if the lines RS, VT meet each other in the point Q of the afymptote CY, and the hyperbola or oppofite hyperbolas in the points R, S and V, T, and if they be parallel to the two femidiameters CA, CB; by Prop. XXXVIII. XXXIX. the rectangle RQS will be equal to the fquare of the femidiameter CA, and the rectangle VQT, or the fquare of the tangent QVT will be equal to the fquare of the femidiameter CB: therefore the rectangle RQS is to the rectangle VQT or to the fquare of the tangent QVT as the fquare of CA to the fquare of CB. Then if two lines parallel to RS, VT meet each other in a point not in the afymptote; the fquares of thefe lines, or the rectangles under their fegments, will be to each other as the fquares or rectangles under the fegments of RS, VT (by 18. of this Book) and therefore are as the fquares of the femidiameters CA, CB to which thefe lines are parallel. Fig. 33.

Cor. 1. Hence, if from any point P two right lines PO, PX be drawn touching the fame hyperbola or oppofite hyperbolas; the lines PO, PX will be to each other as the femidiameters to which they are parallel; for their fquares are as the fquares of thefe femidiameters. Fig. 34.

Cor. 2. If two right lines MN, TR touching an hyperbola or oppofite hyperbolas in the points O, X meet the afymptotes in M, N and T, R; the right lines MR, TN which join the points of con-

courfe

course will be parallel to each other, and to the line OX joining the points of contact. For (by Prop. XXXVIII. and preced. Cor.) PO will be to PX as OM to XR; therefore OX is parallel to MR (2. 6.) and in like manner because PO is to PX as NO is to TX, by permutation, PO is to NO as PX is to TX, and, by conversion, PO is to PN as PX to PT; therefore the triangles OPX, NPT, having the sides about equal angles proportional, are equiangular (6. 6.) consequently the angle TNP is equal to the alternate angle XOP, and therefore the line XO is parallel to TN.

Cor. 3. Hence (the rest remaining as premised) the segment NR of the asymptote CZ intercepted by two right lines touching the hyperbola or opposite hyperbolas, will be bisected by the right line OX (produced if necessary) joining the points of contact; for OX is parallel to MR (by preced. Cor.) let it meet then (if necessary produced) the asymptote CZ in B, and because MO, ON are equal (36. of this Book) the lines RB, BN will be equal.

PROP. XLI.

Fig. 35. **If a right line cutting an hyperbola or opposite hyperbolas in the points E, F, or touching the hyperbola in O, meets a second diameter MN in the point P; the rectangle EPF under the segments of the secant, or the square of the tangent PO, will be to the square of the segment PC between the right line and center, together with the square of the semidiameter MC, as the square of the semidiameter CA, to which the secant or tangent is parallel, to the square of the semidiameter MC which the secant or tangent meets.**

THROUGH the point P let the right line QR be drawn ordinately applied to MN, and let LV be the conjugate diameter to MN and therefore parallel to the line QR; by the preceding Prop. the rectangle FPE will be to the rectangle (QPR or) to

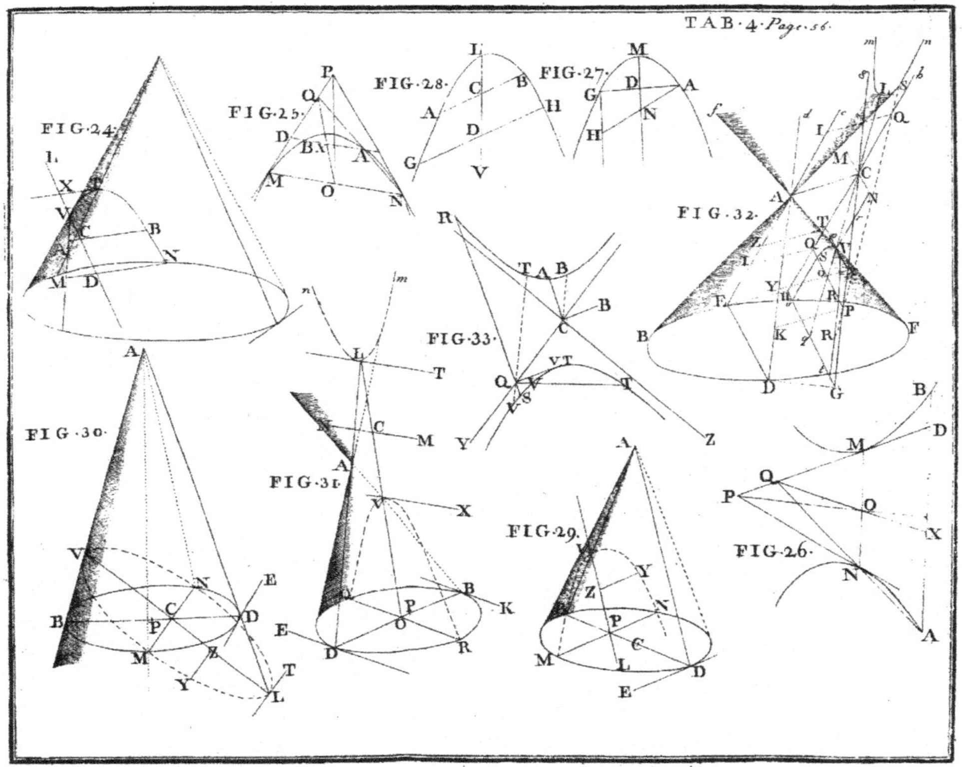
TAB·4·Page·56·
FIG·24·
FIG·25·
FIG·28·
FIG·27·
FIG·32·
FIG·33·
FIG·30·
FIG·31·
FIG·29·
FIG·26·

to the ſquare of PQ as the ſquare of CA to the ſquare of CL, and by Prop. XXXII. the ſquare of QP is to the ſquares of PC, MC together, as the ſquare of CL to the ſquare of CM: therefore, ex æquo, the rectangle FPE will be to the ſquares of PC, CM together, as the ſquare of CA to the ſquare of CM. In like manner the ſquare of the tangent PO will be to the ſquares of PC, CM together, as the ſquare of CA to the ſquare of CM. *Q. E. D.*

PROP. XLII.

If two right lines be parallel to each other, and if both touch or cut, or one of them touches and the other cuts an hyperbola or oppoſite hyperbolas, and meet a right line parallel to one of the aſymptotes; the ſquares of their ſegments between the line parallel to the aſymptote and the points of contact, or the rectangles under their ſegments between this line and the hyperbola or oppoſite hyperbolas, will be to each other as the ſegments of that line parallel to the aſymptote, intercepted between the parallels, and the point in which that line meets one of the hyperbolas.

FOR (by Cor. 1. and 3. Def. XXV.) a line parallel to an aſymptote is parallel to a ſide of the conical ſurfaces in which the hyperbolas are, and meets one of the hyperbolas in one point only; therefore this Prop. is evident, from Prop. XIX. and Cor. 1. to Def. of the Sections.

Cor. 1. Hence if the tranſverſe diameter LV of an hyperbola meets a line KH in O, parallel to the aſymptote CY, meeting the other aſymptote CZ in P, and the hyperbola in E; the rectangle VOL under the ſegments of the diameter between its two vertices and the line KH will be to the ſquare of the ſemidiameter CV as the ſegment OE of the line KH between the diameter and hyperbola to the ſegment EP of the ſame line between the hyperbola and aſymptote CZ; for through P draw a line parallel to the diameter Fig. 36.

LV, meeting the hyperbolas in the points R, S: then by this Prop. the rectangle VOL will be to the rectangle RPS or (Prop. XXXIX.) to the square of CV as OE to EP.

Cor. 2. Or (the rest remaining the same) if the second diameter MN meets KH in B; the square of the segment CB of the diameter between the center and the line KH, together with the square of the semidiameter CN, will be to the square of the semidiameter CN as the segment BE of the line KH between the diameter and hyperbola, to the segment PE of the same line between the asymptote CZ and hyperbola.

For through the points B, P draw the right lines AD, RS ordinates to the diameter MN, and let LV be the diameter conjugate to MN, and consequently parallel to its ordinates; then (by Prop. XXXII.) the square of BC together with the square of CN, will be to the square of CN as the square of AB to the square of CV; that is as the rectangle ABD to the rectangle RPS (39. of this Book) that is (by this Prop.) as BE to PE.

PROP. XLIII.

If any two right lines be drawn from a point in an hyperbola to the asymptotes, and two other right lines be drawn from any other point in the same or opposite hyperbola to the asymptotes and parallel to the two first drawn lines; the rectangle under the two first lines will be equal to the rectangle under the other two lines.

Fig. 37. LET A, B be the points in the hyperbola or opposite hyperbolas, and through A draw the right lines AD, AE to the asymptotes, and through B the lines BG, BF parallel to them; the rectangle under AD, AE will be equal to the rectangle under BG, BF.

For draw the right line AB meeting the asymptotes in the points M, N, and (by Prop. XXXVI.) the segments AM, BN are equal, and therefore AN, BM will be equal; and because the triangles ADM, BGM are equiangular, AD is to BG as AM to BM, or as BN to AN, that is, because the triangles BNF, ANE are equiangular

angular as BF to AE; therefore AD is to BG as BF to AE, and therefore the rectangles DAE, FBG will be equal. *Q. E. D.*

Cor. 1. Hence if two right lines AD, BG be drawn from two points A, B in an hyperbola or opposite hyperbolas, to the same or different asymptote and parallel to the other asymptote; the rectangle under the line AD and the abscis DC between this line and the center C, will be equal to the rectangle under the line BG and abscis GC; for complete the parallelograms CDAE, CGBF, the rectangles DAE, FBG, that is, the rectangles ADC and BGC will be equal.

Cor. 2. And (the same thing being supposed) because the rectangles DAE, FBG are equal; DA is to BG as BF to AE; but the parallelograms DAEC, FBGC are equiangular, and therefore they will be equal (14. 6.)

Cor. 3. If a line AD be drawn from any point A in an hyperbola AVB to one of the asymptotes and parallel to the other, and if a point P in either asymptote be taken, from which a line PO is drawn towards the adjacent hyperbola and parallel to the other asymptote, and if the rectangle CPO be equal to the rectangle CDA; the point O will be in the hyperbola towards which the line PO was drawn: for PO meets somewhere the adjacent hyperbola (Cor. 3. to Def. XXV.) and if it does not meet it in O, a rectangle (Cor. 1.) greater or less than CPO would be equal to the rectangle CDA, contrary to the hypothesis.

DEFINITION XXVI.

LET LV be a transverse diameter of two opposite hyperbolas, and MN the second diameter conjugate to LV; and by Prop. XXXV. let KMS, HNX be two opposite hyperbolas, of which let MN be a transverse diameter and LV the second diameter conjugate to MN; these four hyperbolas are called *conjugate hyperbolas.* Fig. 38.

Cor. 1. It is evident that the point C, the interſection of the diameters LV and MN, is the common center of the conjugate hyperbolas, and that the aſymptotes are the ſame (for the things remaining which were aſſumed in the Definition) let VM be joined, and draw CY biſecting VM in D, and CZ parallel to VM; becauſe LV, MN are conjugate diameters of both hyperbolas TVO, KMS, the lines CY, CZ are aſymptotes of the hyperbola TVO its oppoſite hyperbola, and likewiſe of the hyperbola KMS conjugate to TVO, and its oppoſite HNX (by Cor. 2. 38. of this Book.)

Cor. 2. Becauſe conjugate hyperbolas have the ſame aſymptotes; it is evident from Prop. XXXVI. that if a right line terminated by oppoſite hyperbolas, touches an hyperbola conjugate to them; it will be biſected in the point of contact; or if it meets a conjugate hyperbola in two points; the ſegments between this hyperbola and the oppoſite hyperbolas will be equal.

PROP. XLIV.

If a ſecond diameter of any hyperbola be drawn; its two vertices will be in the hyperbolas conjugate to that hyperbola.

Fig. 38. LET LV, MN be the common conjugate diameters of any conjugate hyperbolas, as in the preceding Definition, and CY, CZ their common aſymptotes, and join VM meeting the aſymptote CY in D, and draw a ſecond diameter BC*b* of the hyperbola TVO; and the two vertices in B, *b*, will be in the hyperbolas conjugate to TVO.

For draw the ſecond diameter CA conjugate to CB, and join AB meeting the aſymptote CY in E; it will be biſected in E, and be parallel to the other aſymptote CZ (Cor. 1. Prop. XXXVIII.) and conſequently parallel to MV; becauſe BE, EA are equal, the rectangle BEC will be equal to the rectangle AEC, that is to VDC (by (by Cor. 1. Prop. preced.) or to the rectangle MDC, becauſe VD, DM are equal; therefore ſince the point M is in the hyperbola KMS and the lines MD, BE are parallel to the aſymptote CZ, and the rectangles MDC, BEC are equal; the point B will be in the ſame hyperbola KMS (Cor. 3. Prop. preced.) and its other vertex *b* in the hyperbola HNX oppoſite to KMS.

Cor. 1. Hence, if right lines be drawn through the vertices of two conjugate diameters CA, CB and touch conjugate hyperbolas, they will meet each other in the aſymptote CY, which the right line joining the vertices A, B meets; for through the point A draw a right line touching the hyperbola TVO, and meeting the aſymptote CZ in R, and CY in P, and join PB meeting the aſymptote CZ in Q; then becauſe PAR touches the hyperbola, the lines PA, AR are equal (36. of this Book) therefore becauſe RQ, AB (Cor. 1. Prop. XXXVIII.) are parallel, PB, BQ will be equal; therefore the line PBQ touches the hyperbola in B (by Prop. XXXVI.) and conſequently the tangents drawn through A, B, meet in the point P in the aſymptote CY.

Cor. 2. Hence likewiſe, any ſecond diameter of an hyperbola is the tranſverſe diameter of its conjugate hyperbola; and the contrary.

PROP. XLV.

If two right lines be drawn from two points in conjugate hyperbolas to the ſame, or different aſymptotes, and parallel to the aſymptotes; the rectangles contained by theſe lines and the abſciſſes, between theſe lines and the center, will be equal.

LET CZ, CY be the aſymptotes to the conjugate hyperbolas, and draw, from a point A in one of them, AE to the aſymptote CY, and parallel to CZ, and let FG be drawn from a point F in a conjugate hyperbola to one aſymptote and parallel to the other; the rectangles AEC, FGC will be equal. Fig. 38.

Let

Let the vertices V, M of the conjugate diameters be joined, the right line joining these points will be parallel to one of the asymptotes and will be bisected by the other in D, and (by Cor. 1. Prop. XLIII.) the rectangles AEC, VDC will be equal, and likewise the rectangles FGC, MDC; but the rectangles VDC, MDC are equal, and therefore the rectangles AEC, FGC are equal. *Q. E. D.*

Cor. 1. Hence, the parallelograms AEC*e*, FGC*g* will be equal, as in Cor. 2. Prop. XLIII.

Cor. 2. Hence also if any two right lines PR, *pr* touch the same hyperbola, or opposite, or adjacent hyperbolas, and meet the asymptotes in the points P, R and *p*, *r*; the triangles PCR, *p*C*r* contained between the tangents and the segments of the asymptotes will be equal. For from the points A, F of contact, draw AE, A*e* and FG, F*g* parallel to the asymptotes; then because PA, AR are equal, PE, EC, and C*e*, *e*R will be equal: therefore the triangle PCR is double the parallelogram AEC*e*: in the same manner it may be shewn that the triangle *p*C*r* is double the parallelogram FGC*g*: but these parallelograms are equal (Cor. preced.) therefore the triangles are equal.

Cor. 3. Hence also, if a right line terminated by adjacent hyperbolas be parallel to one of the asymptotes, it will be bisected by the other asymptote. For let the line AB be terminated by the adjacent hyperbolas and be parallel to the asymptote CZ, and let it meet CY in E; because by this Prop. the rectangles AEC, BEC are equal, the lines AE, EB will be equal, as is evident.

Cor. 4. And on the contrary, if AB, terminated by the conjugate hyperbolas, be bisected by the asymptote CY; it will be parallel to the other asymptote CZ; for, if not, a right line may be drawn from the point A to another point K in the adjacent hyperbola, parallel to CZ: this line AK will also be bisected by the asymptote CY (by preced. Cor.) therefore the asymptote CY will be parallel to the line joining the two points B, K in the hyperbola (2. 6.) which is impossible (Cor. 3. to Def. XXV.)

Cor.

Cor. 5. It appears from this Prop. and Cor. 2. Prop. XLIII. that if two right lines be drawn from points in the same hyperbola, or from points in any two of the four conjugate hyperbolas to one asymptote and parallel to the other; these lines will be reciprocally as the segments of the asymptotes between these lines and the center; and consequently the hyperbola and asymptotes produced indefinitely continually approach each other, and arrive at an interval less than any given distance.

Cor. 6. If through the point A in the hyperbola TAO, a right line be drawn meeting the conjugate hyperbolas BMS, HNX in the points K and H; the rectangle KAH will be double the square of the semidiameter CM which is parallel to KH; for through the point A draw the semidiameter CA, it will be a second semidiameter of the hyperbolas BMS, HNX (Cor. 2. 44. of this Book) therefore by Prop. XLI. the rectangle KAH will be to twice the square of CA as the square of CM to the square of CA; therefore the rectangle KAH is double the square of CM.

PROP. XLVI.

If through a point in an hyperbola two indefinite right lines be drawn parallel to the asymptotes and meeting any diameter; the semidiameter will be a mean proportional between the segments of this diameter intercepted between the center and the lines which it meets.

DRAW through the point A of the hyperbola TVO, the right lines DE, FG parallel to the asymptote CY, CZ, and first let them meet the transverse diameter LCV in the points P, Q; CP, CV, CQ will be proportional. Fig. 39.

For let FG meet the asymptote CY in X, and from the points V, P draw to CY the lines V*r*, PS parallel to FG or CZ; PS will be equal to AX; and because V*r*, PS are parallel, CS will be to C*r*, as (SP or) XA to *r*V, that is (by Cor. 5. preced.) as C*r* to CX; therefore because CS, C*r*, CX are proportional, CP, CV, CQ will be proportional, for the lines SP, *r*V, XQ are parallel.

Now

Now let the lines DE, FG meet the second diameter MCN in the points K; H, and let its vertex M be in the hyperbola conjugate to TVO, which let the line FG meet in the point B, through which point draw BR parallel to DE or CY, meeting the diameter MN in R; by the first case, CR, CM, CH are proportional: but because BR, XC, DE are parallel, and the lines AX, XB are equal (by Cor. 3. preced.) the lines KC, CR will be equal; therefore CK, CM, CH are proportional. *Q. E. D.*

S C H O L I U M.

FIG. 39. *IF two indefinite right lines DE, FG cut each other in the point A, and another indefinite line revolves about any given point C taken without the lines DE, FG, and in this revolving line, whilst it meets (as in the points P, Q) the sides of the angle EAF (without which is assumed the point C) or the sides of its vertical angle GAD, let a point V be taken on the same side of the point C with the points P, Q, so that the segments CP, CV, CQ be always proportional; the* Locus *of all the points V will be the hyperbola TVO passing through the point A, whose center is the given point C, and its asymptotes will be parallel to DE, FG, and the revolving line will at length coincide with the asymptotes; and if on the contrary side of the point C, the point L be so taken that CL be always equal to CV; the hyperbola opposite to TVO will be the* Locus *of the points L. But if in the revolving line, whilst it meets (as in the points K, H) the sides of the angle EAG within which the point C is placed, two points M, N be so taken, that the segments CK, CM or CN, and CH be proportional, the* Loci *of all the points M, N will be the opposite hyperbolas conjugate to the former.*

PROP.

PROP. XLVII.

If from a point F in an hyperbola a right line FE be drawn ordinately applied to a diameter BA, and through the ſame point F a tangent be drawn meeting this diameter in D; the ſegment of the diameter between the ordinate and tangent will be biſected in its vertex B. FIG. 40.

FOR let the ordinate FE produced meet the parabola again in L, and draw through the point F the diameter FG, and the lines LG, EC, BH parallel to the tangent FD; LG, BH will be ordinates to the diameter FG (Cor. 27. of this Book) and EC will be equal to BH; then becauſe LF is double EF, GF will be double CF and LG double EC or BH (2. 6.) and therefore the ſquare of LG is quadruple the ſquare of BH (4. 2.) and therefore the abſciſs GF is quadruple the abſciſs HF (29. of this Book) and therefore CF is double HF, and conſequently ED double BD; and therefore the ſegment ED of the diameter is biſected at the vertex B. *Q. E. D.*

PROP. XLVIII.

If from a point F in an ellipſe an ordinate FE be drawn to the diameter AB, and alſo a right line touching the ellipſe and meeting the ſame diameter in D; the ſemidiameter CB will be a mean proportional between the ſegments CE, CD of the diameter between the center and ordinate, and between the center and tangent. FIG. 41.

FOR draw through the two vertices of the diameter AB right lines touching the ellipſe, and meeting the tangent drawn through F in the points H, G; theſe tangents will be parallel to the ordinate FE (Cor. 1. 25. of this Book); take CT in the diameter AB equal to CE; AT, EB will be alſo equal: but (by Cor. 5. Prop. XVIII. of this Book, and 22. 6.) AH is to BG as HF to FG or (becaufe HA, FE, GB are parallel lines) as AE to EB; therefore AE is to EB as AD to BD; and therefore by divifion, TE will be to EB as AB to BD, and by taking the halves of the antecedents CE will be to EB as CB to BD, and therefore by compofition and converfion, CE is to CB as CB to CD. *Q. E. D.*

PROP. XLIX.

If from a point in an hyperbola an ordinate be drawn to a diameter, and alfo a right line touching the hyperbola and meeting the fame diameter; the femidiameter will be a mean proportional between the fegments of the diameter between the center and ordinate and between the center and tangent.

FIG. 42. *Cafe* 1. WHEN an ordinate is drawn to a tranfverfe diameter: let AB be the tranfverfe diameter of the hyperbola, C the center, and from the point F in it draw an ordinate FE applied to the diameter AB, and a tangent meeting the fame diameter in D; CE will be to CB as CB to CD.

Draw through the two vertices of the diameter AB two tangents meeting the tangent drawn through F in the points H, G, which will be parallel to the ordinate FE (Cor. 1. Prop. XXV. of this Book) and take CT equal to CE, and AT, BE will be alfo equal: but (by Cor. 5. Prop. XVIII. of this Book, and 22. 6.) AH is to BG as HF to GF, that is (becaufe AH, GB, FE are parallel) as AE to BE, and as AH is to BG fo is AD to BD; therefore AE will be to BE as AD to DB, and by compofition TE will be to BE as AB to DB, and by taking the halves of the antecedents, CE is to BE as CB to DB, and therefore by converfion CE is to CB as CB to CD.

Cafe 2. When an ordinate is drawn to a fecond diameter. From the point F in an hyperbola draw FO an ordinate to the fecond diameter MN, which the tangent drawn through F meets in P, and let AB be the tranfverfe diameter conjugate to MN, meeting the tangent

gent FP in D, and draw FE ordinately applied to AB; it will be parallel and equal to CO.

Becaufe, by the firft cafe, EC, BC, DC are proportional, the fquare of EC will be to the fquare of BC as EC to DC (Cor. 20. 6.) therefore by divifion the rectangle BEA is to the fquare of BC as ED to DC; but the fquare of the ordinate FE is to the fquare of the fecond femidiameter CM as the rectangle BEA to the fquare of BC, as appears from Def. XXIV. therefore the fquare of FE is to the fquare of CM (as ED to DC) that is as FE to CP; therefore FE, CM, CP are proportional (by the converfe of Cor. 20. 6.) and therefore CO, CM, CP are proportional. *Q. E. D.*

Corollaries to the two preceding Propofitions.

Cor. 1. If a right line touches an ellipfe or hyperbola in any point F, and meets the diameter AB (which in the hyperbola fhould be a tranfverfe diameter) in the point D, and if an ordinate FE be drawn from the point of contact to the diameter; the rectangle CED under the fegments of the diameter, between the ordinate and center, and between the fame and the tangent, will be equal to the rectangle AEB under the fegments between the ordinate and the two vertices of the diameter. Fig. 41, 42.

For becaufe (by the preced. Propofitions, and 17. 6.) the rectangle ECD is equal to the fquare of CB, if in the ellipfe the fquare of CE be taken from each of them; the remainders, viz. the rectangles CED, and AEB (by the 3. and 5. 2.) will be equal, and in the hyperbola thefe equals, viz. the rectangle ECD and the fquare of BC being taken from the fquare of EC, the remainders viz. the rectangles CED and AEB (by 2. and 6. 2.) will be equal.

Cor. 2. Let the tangent and ordinate drawn from the point F, meet the fecond diameter CM in P and O, and becaufe the rectangle PCO is equal to the fquare of CM; let the fquare of CO be added to each, and the rectangle POC (3. 5.) will be equal to the fquares of CM, CO together. Fig. 42.

Fig. 41, 42. Cor. 3. And the rectangle ADB, under the segments between the tangent and the two vertices of the diameter (which in the hyperbola should be the transverse) will be equal to the rectangle CDE under the segments between the tangent and center and between the same tangent and ordinate: for in the ellipse, equals being subducted, viz. the square of CB and the rectangle ECD from the square of CD, the remainders, viz. ADB and CDE (by 6. and 2. 2.) will be equal, and in the hyperbola, if from equals viz. from the square of CB and from the rectangle ECD, the square of CD be taken, the remainders ADB and CDE will be equal (5. and 3. 2.)

PROP. L.

If two parallel right lines touching an ellipse or opposite hyperbolas, meet any other tangent; the rectangle under the segments of the parallel lines between the points of contact and the tangent which they meet, will be equal to the square of the semidiameter, to which the tangents are parallel: and the rectangle under the segments of the tangent which the parallels meet, viz. between its point of contact and the parallel lines, will be equal to the square of the semidiameter parallel to this tangent.

Fig. 41, 42. Let the parallel lines AH, BG touch an ellipse or opposite hyperbolas in the points A, B, and let them meet a tangent drawn through any point F in H and G; the rectangle under AH, BG will be equal to the square of the semidiameter CM parallel to AH, BG, and the rectangle HFG will be equal to the square of the semidiameter CK parallel to HG.

For the line joining the points A, B will be a diameter (Cor. 2. 22. of this Book) and is conjugate to CM (Cor. 27. of this Book) if the tangent HG in the ellipse be parallel to AB, the proposition is manifest; but if it be not parallel to AB, let it meet it in D and CM produced in P, and from the point F draw FE, FO ordinately applied to the diameters AB, CM.

Part

Part 1. Becaufe by Cor. 3. preced. the rectangles ADB and CDE are equal; AD will be to CD as ED to BD (16. 6.) therefore (being parallel lines) AH is to CP as EF to BG (2. 6.) and therefore the rectangle under AH, BG is equal to the rectangle under CP, EF or CO, that is, to the fquare of the femidiameter CM: for by the two preceding Propofitions, CO, CM, CP are proportional.

Part 2. Becaufe AH is to BG as HF to FG (Cor. 5. 18. of this Book, and 22. 6.) the rectangle HFG is fimilar to the rectangle under AH, BG and therefore is to the fame rectangle AH, BG as the fquare of HF to the fquare of AH, that is, as the fquare of CK to the fquare of CM (by 31. and 40. of this Book) therefore by permutation, the rectangle HFG is to the fquare of CK as the rectangle AH, BG to the fquare of CM, that is, in a ratio of equality by the firft part. *Q. E. D.*

PROP. LI.

If a right line touching an ellipfe or hyperbola in any point F, meets two conjugate diameters AB, CM in the points D, P; the rectangle PFD under the fegments of the tangent between the point of contact and the diameters will be equal to the fquare of the femidiameter CK parallel to the tangent. Fig. 41, 42.

FOR draw from the point of contact F a right line FE ordinately applied to the diameter AB (which in the hyperbola fhould be the tranfverfe diameter) and draw through the two vertices of this diameter two tangents, and let them meet the tangent drawn through F in the points H, G; they will be parallel to the ordinate FE and to the diameter CM; then becaufe (by Cor. 1. 49. of this Book) the rectangle CED is equal to the rectangle AEB, CE is to EB as AE to ED; therefore (becaufe of the parallel lines) PF is to FG as HF to FD, and therefore the rectangle PFD is equal

equal to the rectangle HFG, that is, to the ſquare of the ſemidiameter CK, by the ſecond part of the preced. Prop. *Q. E. D.*

Cor. 1. It is manifeſt, if the right line PFD touching the ellipſe or hyperbola, meets two diameters CB, CM, and if the rectangle PFD be equal to the ſquare of the ſemidiameter CK conjugate to that which paſſes through the point of contact F, that CB, CM are conjugate diameters.

PROP. LII. PROBL. IV.

A conic ſection being given in poſition, and a point given without the ſection; to draw two right lines which ſhall touch the ſection or oppoſite ſections, provided the given point be not in an aſymptote of an hyperbola.

Fig. 40, 41, 42, 43. LET FL be the given ſection, and D the point given; draw through the point D a diameter meeting the ſection in B, or if it be a ſecond diameter of an hyperbola let its vertex or extremity be B: or if the ſection be a parabola take BE equal to BD, and through E draw FEL parallel to the ordinates applied to the diameter BE, meeting the parabola in the points L, F; the lines joining the points, DL, DF touch the ſection; for if either of them, ſuppoſe DL, does not touch the ſection, let LR be a tangent meeting the diameter BE in R, (by the 47. of this Book) BR is equal to BE, that is to BD, which is abſurd.

But if the ſection be an ellipſe or an hyperbola; let C be its center, and let CE be a third proportional to CD, and CB (which muſt be taken on the ſide of the center, oppoſite to that on which CD is placed, if CD be a ſecond diameter of the hyperbola) but if not, take it on the ſame ſide, and through the point E draw FEL parallel to the ordinates applied to the diameter CD, meeting the ſection in the points L, F; the lines joining theſe points DL, DF touch the ſection or oppoſite ſections.

For if either of them, ſuppoſe DL does not touch the ſection, let

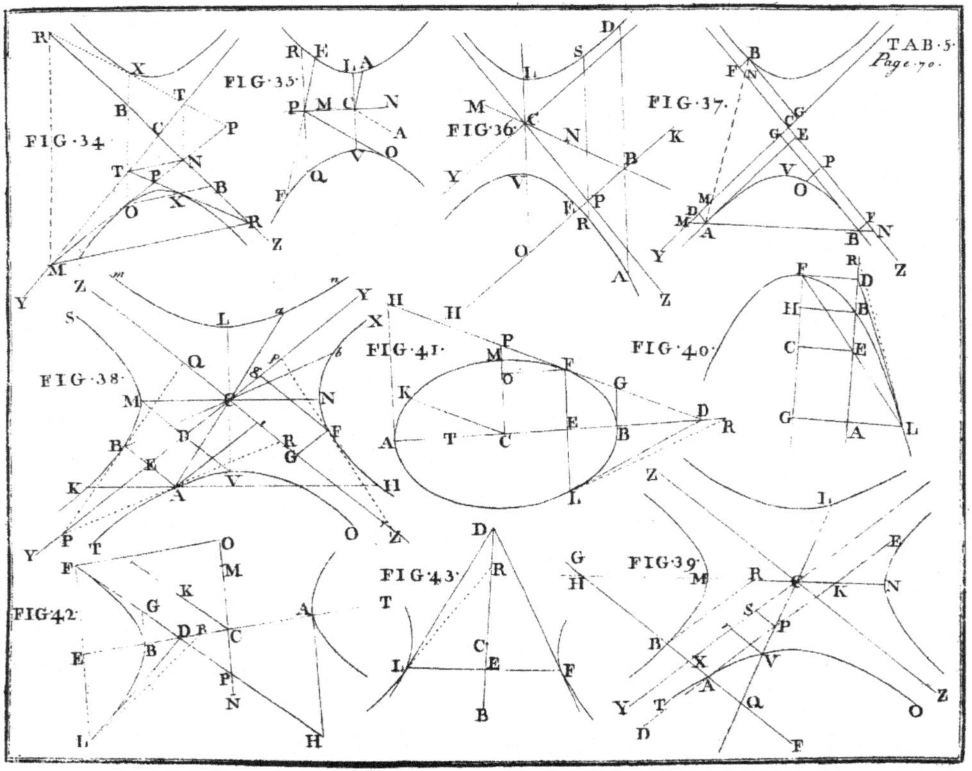
TAB. 5.
Page 70.
FIG. 34.
FIG. 35.
FIG. 36.
FIG. 37.
FIG. 38.
FIG. 41.
FIG. 40.
FIG. 42.
FIG. 43.
FIG. 39.

let LR be a tangent meeting the diameter CD in R; the lines CE, CB, CR will be proportional (by the 48. 49. of this Book) which is absurd.

If the given point D be in one of the asymptotes of an hyperbola, two tangents cannot meet in this point (Cor. 4. 39. of this Book) but a method of drawing a right line which shall touch the adjacent hyperbola was shewn in Cor. 5. Prop. XXXVI.

PROP. LIII.

If two right lines touching a conic section or opposite sections meet each other, and through a point in one tangent a right line be drawn parallel to the other, and meeting the right line joining their points of contact and also the conic section in two points; the square of its segment between the tangent and the line joining the points of contact will be equal to the rectangle under the segments intercepted between the tangent and section. Fig. 44, 45.

LET the lines AB, AC touch a conic section or opposite sections in the points B, C, and join these points and draw a right line through the point D in the tangent AC parallel to the other tangent AB, meeting BC in G and the section in the points E, F; the square of DG will be equal to the rectangle EDF.

For because the tangent AC meets the parallels FD, BA, the rectangle EDF under the segments of the secant is to the square of the tangent BA as the square of DC to the square of AC (by Cor. 6. 18. of this Book) that is (from similar triangles) as the square of DG to the square of BA; therefore the rectangle EDF is equal to the square of DG. *Q. E. D.*

Cor. Let two right lines touching a conic section, or opposite sections meet each other in the point A, if through a point P in the tangent AC a right line be drawn parallel to the other tangent, and meeting the right line joining their points of contact in L, and if any other line be drawn from the same point P cutting the section

tion or opposite sections in the points R, Q; the rectangle RPQ and the square of PL will be to each other as the squares of the segments of the tangents, which are parallel to PRQ, and PL, intercepted between their points of concourse and contact; or as the squares of the semidiameters parallel to PRQ and PL.

For through the point D in the tangent AC draw two right lines parallel to PRQ, PL, cutting the section or opposite sections in the points O, N and E, F, and let the right line DEF parallel to PL meet the right line joining the points of contact in G: then because the tangent AC meets two parallel secants (Cor. 4. 18. of this Book) the rectangle RPQ will be to the rectangle ODN as the square of PC to the square of DC, that is from similar triangles as the square of PL to the square of DG or to the rectangle EDF (by this Prop.) therefore by permutation the rectangle RPQ is to the square of PL as the rectangle ODN to the rectangle EDF, that is (by the 18. of this Book) as the squares of the segments of the tangents which are parallel to PRQ, PL, intercepted between their points of concourse and contact: or as the squares of the semidiameters of the ellipse or hyperbola which are parallel to PRQ, PL, by (31. and 40. of this Book).

LEMMA II.

FIG. 46. FROM a point E in any indefinite right line take the lines EA, ED on the same side of the point E, and from the same point E take the lines EB, that its extremity B be either between the points A, D, or on the contrary side of the point E; if the square of AB be to the square of DB as AE to DE, EB will be a mean proportional between the lines AE and DE.

For if not, take from the point E towards B a line EF (greater or less than EB) a mean proportional between AE and DE; then because AE, FE, DE are proportional, by conversion and permutation, as AE the first is to FE the second term, so will AF the sum or difference of the first and second terms be to DF the sum or differ-

difference of the ſecond and third; therefore the ſquare of AF will be to the ſquare of DF as the ſquare of AE to the ſquare of FE, that is as AE to the third DE, or, by conſtruction, as the ſquare of AB to the ſquare of DB; therefore AF is to DF as AB to DB, which is abſurd; as is ſelf-manifeſt when the points F, B are between A, D, and if the points F and B be on contrary ſides of the point E, it is evident alſo that the ratio of the unequal lines AF, DF will not remain the ſame, if to theſe unequal lines, the ſame line FB be added or taken from them; therefore no line either greater or leſs than EB is a mean proportional between AD, DE; and therefore EB is a mean proportional between theſe right lines.

PROP. LIV.

If a diameter of a parabola, or a right line parallel to one of the aſymptotes of an hyperbola, meets two tangents and the right line joining the points of contact; the ſquare of its ſegment between the ſection and the line joining the points of contact, will be equal to the rectangle under its other ſegments, viz. between the ſection and tangents.

LET two lines MN, MO touch a parabola, an hyperbola or op- FIG. 47.
poſite hyperbolas in the points N, O; draw the line TX, which 48.
may either be a diameter of the parabola or be parallel to one of the aſymptotes of the hyperbola, meeting the parabola or either of the hyperbolas in E, the tangents in A, D and the line joining the points of contact in B; the ſquare of EB will be equal to the rectangle AED.

Caſe 1. If the lines touching the parabola or hyperbola meet FIG. 47.
each other in M, through the point D, in which the line TX meets one of the tangents, let a line be drawn parallel to the other tangent, cutting the ſection in the points P, L and the line joining the points of contact in K, which may be always done; then by the preced. Prop. the ſquare of DK is equal to the rectangle PDL, and becauſe AO is parallel to DK, the ſquare of AB will be to the

ſquare of DB as the ſquare of AO to the ſquare of DK, or to the rectangle PDL, that is (Prop. XIX.) as AE to DE; but the lines AE, BE, DE are taken from the point E as in the preceding Lemma; therefore they are proportional by the ſame Lemma, and conſequently the ſquare of BE is equal to the rectangle AED.

FIG. 48. *Caſe* 2. If the right lines touching the oppoſite hyperbolas meet each other in M; through either of the points A or D, ſuppoſe D, draw a right line parallel to the other tangent AO and meeting the line joining the points of contact in K, and let two right lines be drawn through the points A, D parallel to each other, cutting the oppoſite hyperbolas in the points G, H and P, L: by Prop. XL. the rectangle GAH and the ſquare of AO will be to each other as the ſquares of the ſemidiameters parallel to the lines GH, AO, and by Cor. preced. the rectangle PDL will be to the ſquare of DK as the ſquares of the ſame ſemidiameters; therefore, by permutation, the rectangle GAH will be to the rectangle PDL as the ſquare of AO to the ſquare of DK or (becauſe AO, DK are parallel) as the ſquare of AB to the ſquare of DB; but (Prop. XIX.) the rectangle GAH is to the rectangle PDL, as AE to DE and conſequently the ſquare of AB is to the ſquare of DB as AE to DE; therefore the lines AE, BE, DE (by Lemma preced.) are proportional, and therefore the ſquare of BE is equal to the rectangle AED.

Caſe 3. But if the lines AO, DN touching the oppoſite hyperbolas ſhould be parallel to each other, then their ſquares would be to each other as the ſquares of AB, DB, becauſe the triangles ABO, DBN are ſimilar, and they would be likewiſe as AE, DE by Prop. XIX. therefore the ſquare of AB would be to the ſquare of DB as AE to DE, and therefore the lines AE, BE, DE would be proportional and the ſquare of BE equal to the rectangle AED, as before. *Q. E. D.*

COR. 1. If the right line TX be a diameter of a parabola, or parallel to one of the aſymptotes of an hyperbola, and meets the ſection in E, and in the point A the line touching the ſection in O, and in B a line cutting the ſection or oppoſite ſections in the points O, N;

O, N; and if from the point E in TX without the section DE be taken a third proportional to AE, BE; the line joining the points D, N will be a tangent. For if not, a line touching the section in N would meet TX without the section in some point C different from the point D, and therefore by this Prop. CE would be a third proportional to AE, BE, contrary to the hypothesis; therefore the line DN touches the section in N.

Cor. 2. Hence, if, from M the point of concourse of two right lines touching a parabola, an hyperbola, or opposite hyperbolas in the points N, O, a line MF be drawn to the line joining the points of contact, which may be a diameter of the parabola, or be parallel to an asymptote of the hyperbola, it will be bisected by the section in S; for if MS, FS be not equal, let SR in the line MF be taken without the section from the point S, a third proportional to MS, FS and (by Cor. preced.) the line RN or RO joining the points touches the section in N or O (contrary to 16. of this Book) therefore MS, FS are equal.

Prop. XXXV. Book I. and Prop. XXX. XXXI. Book III. of Apollonius, are contained in this Corollary.

SCHOLIUM.

LET the sides of a triangle MNO be indefinitely produced both ways, and let the right line TX cutting all the sides of that triangle be moved in such manner as to be always parallel to itself, and let a point E be so taken in it, that the square of the segment EB between the point E and the given side NO of the triangle be always equal to the rectangle AED under the segments of the line TX between the point E and the other sides of the triangle. If the line TX, when it bisects in its motion the given side NO of the triangle, passes through the opposite angle M; the Locus *of all the points E will be a parabola and its diameters will be parallel to TX, and the lines MN, MO will touch this parabola in the points N, O. But if the line TX in its motion bisecting the side NO, does not at the same time pass through the angle at M; the* Locus *of all the points will be the opposite hyperbolas and one of its asymptotes* Fig. 47, 48.

 will

will be parallel to TX; and that asymptote is easily found, viz. by drawing parallel to TX a right line whose segment intercepted by the sides MN and MO of the triangle would be bisected by the other side NO, as is evident (from Cor. 3. Prop. XL.) when this asymptote passes between the points N, O, the lines MN, MO will touch in the points N, O opposite hyperbolas generated by the motion of the point E. If it does not pass between these points they will touch in those points one of the opposite hyperbolas, generated by the motion of this point E; the right line drawn through M bisecting NO will (by Prop. XXVI.) be a diameter, and its intersection with the asymptote will be the center of the hyperbolas; and from hence the other asymptote is found. Or if the moving line TX meets two parallel lines MO, MN and any other line ON cutting them; and a point E be taken in TX, so that the square of the segment EB of the line TX between the point E and the line ON be equal to the rectangle AED, under the segments between the same point E and the parallels; the point E in its motion will describe the opposite hyperbolas, which the parallels touch in the points O, N.

PROP. LV.

If from two given points in a parabola, an hyperbola or opposite hyperbolas two right lines be inflected to any third point, in the same parabola, or in either of the hyperbolas, and meet a right line given in position, which may be either a diameter of the parabola, or be parallel to an asymptote of the hyperbola; the segments of this line between the inflected lines and the point in which it meets the section, will be to each other always in the same ratio wheresoever in the section the point be taken to which the lines are inflected.

Fig. 49, 50, 51. LET N, M be the two points given in a parabola, or in an hyperbola or opposite hyperbolas, and let TX be a diameter of the parabola, or let it be parallel to one of the asymptotes of the hyperbola meeting the section in the given point E, and from the points N, M to any point O in the parabola, or in either of the hyperbolas,

perbolas, inflect two right lines NO, MO meeting TX in the points B, C; the abscisses EB, EC will be to each other in the same ratio wheresoever the point O be taken in the section.

For draw through the three points M, N, O, three tangents which may meet TX in three points F, D, A, and let MN be joined meeting TX in G; it appears from the preced. Prop. that ED, EG, EF are proportional.

Because the line TX meets the two tangents DN, AO in the points D, A, and the line joining the points of contact in B; the square of EB will be equal to the rectangle AED (by preced. Prop.) and in like manner because the same line meets the tangents MF, AO in the points F, A, and the line joining the points of contact in C; the square of EC will be equal to the rectangle AEF; therefore the square of EB is to the square of EC as the rectangle AED to the rectangle AEF, that is, as ED to EF, or as the square of ED to the square of EG (a mean proportional between ED, EF) therefore EB is to EC as ED to EG; but because the points M, N are given, and the line TX given in position; the segments ED, EG remain always the same; therefore the point O moving through the whole parabola, hyperbola, or the opposite hyperbola, the segments EB, EC of TX between the inflected lines and section will be always in the same ratio, viz. as the lines ED, EG or EG, EF. *Q. E. D.*

This Proposition (so far as it relates to the parabola) is the same which Fermat proposed in his letter to Sir Kenelm Digby to be demonstrated by Dr. Wallis (see page 858. Tom. 2. Oper. Math. Wallisii) but by the demonstration here given, it appears that the same property extends to the hyperbolas also.

PROP.

PROP. LVI.

If a trapezium be inscribed in a conic section or opposite sections, and all its sides be indefinitely produced, and if from any point E in the section two right lines be drawn parallel to two adjacent sides of the trapezium; the rectangles under the segments of these lines between the point in the section and the opposite sides of the trapezium will be to each other, as the squares of the tangents or of the semidiameters to which these lines are parallel.

FIG. 52, 53. *Case* 1. LET AB*d*C be a trapezium inscribed in the section or opposite sections, and first let the two sides AC, B*d* be parallel, and draw from any point E in the section two right lines parallel to the adjacent sides AB, AC meeting the opposite sides of the trapezium in the points Q, *n* and *b*, R; the rectangles QE*n*, *b*ER will be to each other as the squares of the tangents or of the semidiameters to which E*n*, E*b* are parallel. For let *b*ER meet the section again in T, and draw the right line MK bisecting the parallels AC, B*d*, it will be a diameter of the section and will bisect in the point V the line TE terminated by the section (Cor. 6. and 5. Prop. XXV.) and likewise the line *b*R, terminated by the lines C*d*, AB, and parallel to the bisected lines AC, B*d*; therefore *b*E, TR are equal, and therefore the rectangles *b*ER, TRE are equal, and (because of the parallelograms) the rectangles QE*n*, ARB are equal; therefore the rectangles *b*ER, QE*n* are to each other as the rectangles TRE, ARB, that is, as the squares of the tangents or of the semidiameters to which the lines E*b*, E*n* are parallel, by Prop. XVIII. or XXXI. and XL.

Case 2. Let ABDC be a trapezium inscribed, none of whose sides are parallel to each other, and through the point E in the section draw two lines parallel to AB, AC meeting the opposite sides of the trapezium in the points Q, N and H, R; the rectangles HER, QEN

QEN will be to each other as the ſquares of the tangents or of the ſemidiameters to which HE, EN are parallel.

For draw through the point B a right line parallel to the ſide AC, and meeting the ſection again in *d* and the line EN in *n*, let the right line joining the points *d*, C meet EH in *b*, and draw through the point D a right line parallel to the ſide AC, meeting the ſection again in F, and the ſide AB in G, and the line joining the points *d*, C, in S; draw the right line KM biſecting the parallels AC, B*d*; it will be a diameter of the ſection, and will biſect in the point O the line DF terminated by the ſection (as in the former caſe) and likewiſe SG parallel to AC, B*d*; therefore SD and FG are equal: and becauſe the triangles HC*b*, DCS are ſimilar; H*b* will be to (SD or) FG as (C*b* to CS, that is as RA or) EQ to AG; and becauſe the triangles NB*n*, GDB are ſimilar (*n*B or) ER will be to DG as N*n* to BG: therefore by taking the rectangles under the correſponding antecedents of the firſt and ſecond order of the proportionals, and the rectangles under the conſequents of the ſame orders, the rectangle H*b*, ER is to the rectangle DGF, as the rectangle EQ, N*n* to the rectangle AGB; therefore, by permutation, the rectangles under H*b*, ER and EQ, N*n* are to each other as DGF and AGB, that is as the ſquares of the tangents or of the ſemidiameters to which the lines FG and AG, or HE and EN are parallel (by Prop. XVIII. XXXI. and XL.) and by the preceding caſe, the rectangles *b*E, ER, and *n*E, EQ are to each other in this ſame ratio; therefore the ſums or differences of theſe rectangles, that is, the rectangles HER, QEN will be to each other in this ſame ratio. *Q. E. D.*

This Propoſition holds good, although in the firſt caſe the line *b*ER drawn parallel to AC, B*d*, touches the ſection in E; for then the points E, T coinciding, the point E will be at the vertex of the diameter MK of the ſection, and the rectangle *b*ER or TRE becomes the ſquare of *b*E or ER; and in like manner in the ſecond caſe, the points D, F coinciding, the rectangle DGF becomes the ſquare of the tangent DG or DS.

Cor.

Cor. 1. Hence, the points A, B, D, C, remaining fixt; and the point E moving through the section or opposite section, and the lines drawn through it being always parallel to AB, AC, the ratio of the rectangles HER, QEN will be constantly the same.

Cor. 2. Or the points A, B, C, E, and the lines drawn through E remaining fixt as before; the point D (that is the intersection of the lines CD, BD) moving through the section or opposite section, the ratio of the rectangles HER, QEN will continue the same; and therefore (because their sides ER, QE are given) the ratio of the lines HE, NE will always remain the same.

Fig. 52. Cor. 3. A conic section does not meet another conic section or opposite sections in more than four points. For if it be possible, let two sections pass through five points as A, B, C, D, E, and (every thing remaining as assumed above) draw through the point B a right line cutting both sections in the points *d*, P, join C*d* meeting HE in *h*, and join CP meeting the same line HE in L; then because the points *d*, D are in the same section passing through the points A, B, C, E, the lines *h*E, *n*E are to each other as HE, NE (by Cor. preced.) and in like manner when the points P, D are in the same section passing through the points A, B, C, E; the lines LE, *n*E will be to each other as HE, NE; therefore *h*E, LE are equal, which is absurd; for the points *h* and L are on the same side of the point E; from hence the Corollary is manifest.

Fig. 54. Cor. 4. If a trapezium ABCD, whose sides are given in position, be inscribed in a conic section given in position, and if from any point E in the section the right lines EF, EG, EH, EK be drawn in given angles to the four sides of the trapezium, the rectangles FEH, GEK under the lines drawn to the opposite sides, will be in a given ratio to each other.

For draw from the point E the right lines EL, EO parallel to two adjacent sides of the trapezium, viz. to BC, BA; and let them meet the opposite sides in the points L, M, and N, O: then because in the triangle ELF, the angle LFE is given by hypothesis, and

and the angle ELF is equal to the given angle CBA, the triangle ELF (by 40. of Euclid's Data) is given in ſpecies; and therefore the ratio of FE to EL is given. In like manner in the triangle EMH the ratio of EH to EM will be given, and therefore the ratio compounded of theſe ratios is given; that is (by 23. 6.) the ratio of the rectangle FEH to the rectangle LEM.

It may be demonſtrated in the ſame manner that the ratio of the rectangle OEN to the rectangle GEK is given. But the ratio of the rectangle LEM to the rectangle OEN is given (by Cor. 1.) therefore becauſe it has been ſhewn that the ratio of the rectangle FEH to LEM is given, and alſo the ratio of LEM to OEN, and of OEN to the rectangle GEK; the ratio of the rectangle FEH to the rectangle GEK will be given (by 8. Euclid. Dat.)

All the caſes of the 17. Lemma, Book I. of Newton's Math. Principles of Nat. Philoſophy, are fully demonſtrated in the Propoſition and this laſt Corollary.

END OF THE FIRST BOOK.

TAB·6· Page 82

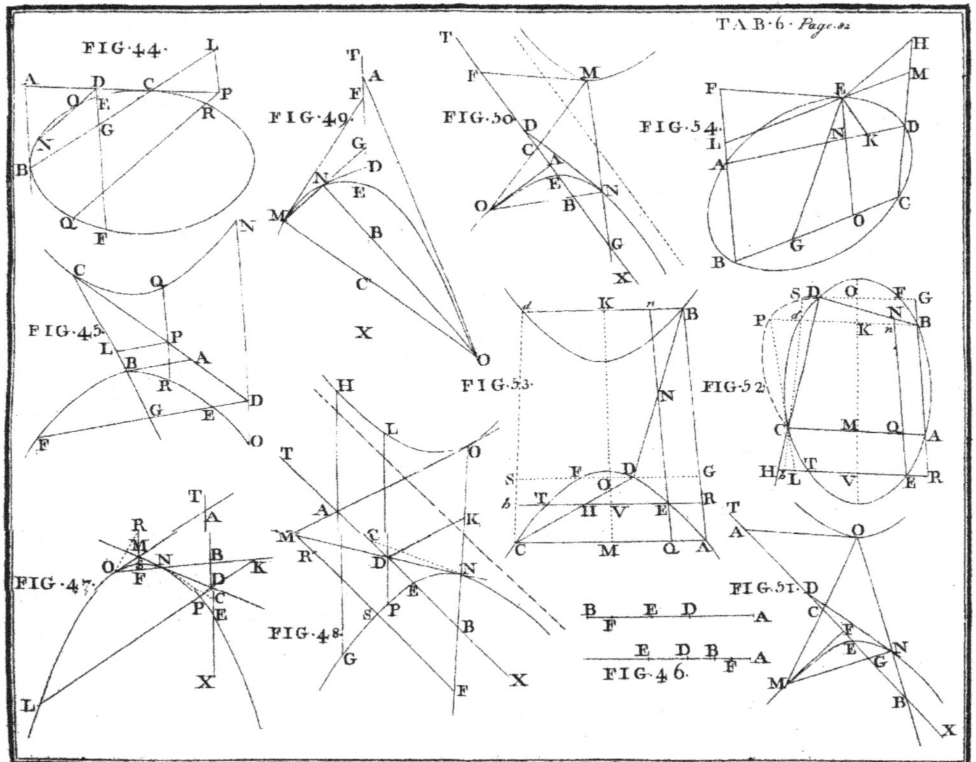
FIG·44·
FIG·45·
FIG·46·
FIG·47·
FIG·48·
FIG·49·
FIG·50·
FIG·51·
FIG·52·
FIG·53·
FIG·54·

CONIC SECTIONS.

BOOK THE SECOND.

Concerning the Properties from which the Sections derive their names; their parameters, axes, foci, and descriptions in plano.

DEFINITIONS I. and II.

I. IF from a point in a parabola a right line be drawn ordinately applied to a diameter; a right line which is a third proportional to the absciss and this ordinate, is called the *Latus Rectum*, or *Parameter* of that diameter.

II. A right line, which is a third proportional to two conjugate diameters of an ellipse or hyperbola, is called the *Latus Rectum* or *Parameter* of that diameter, which is the first of the three proportionals.

L 2 Cor.

Cor. 1. If an ordinate be drawn from a point in an ellipse or hyperbola to a diameter (which in the hyperbola is a transverse diameter) the rectangle under the abscisses of the diameter between its two vertices and the ordinate will be to the square of the ordinate as the diameter to its parameter. For, by Cor. 1. 31. and Def. XXIV. Book I. this rectangle is to the square of the ordinate as the square of the diameter to the square of its conjugate, that is as the diameter to its parameter (by Def. preced. and Cor. 20. 6.)

Cor. 2. If an ordinate be drawn from a point in an hyperbola to a second diameter; the sum of the squares of the second semidiameter and of its segment between the center and ordinate will be to the square of the ordinate as (the square of that second diameter to the square of its conjugate, by Prop. XXXII. Book I. that is as) the second diameter to its parameter, by Def. preced. and Cor. 20. 6.

PROP. I.

The squares of right lines ordinately applied to a diameter of a parabola are equal to the rectangles under the abscisses and its parameter.

Fig. 1. FROM a point B in a parabola let the right line BG be ordinately applied to the diameter FL whose vertex is F, and let the right line P be a third proportional to the abscis GF and ordinate BG; it will be the parameter of the diameter FL, by Def. I. of this Book, and the square of BG will be equal to the rectangle under GF and the parameter P: then if any other right line as DL be ordinately applied to the diameter FL; by Prop. XXIX. Book I. the square of DL will be to the square of BG as LF to GF, that is as the rectangle under LF and P, to the rectangle under GF and P; therefore the square of DL is to the rectangle under LF and P as the square of BG to the rectangle under GF and P, that is, in a ratio of equality, and from hence Apollonius called this section a *Parabola.* *Q. E. D.*

Cor.

Cor. 1. If from a point G in the diameter FL of a parabola, a right line GB be drawn parallel to the ordinates applied to that diameter, and the square of BG be equal to the rectangle under the abscifs GF (between GB and the vertex F of the diameter) and the parameter of the fame diameter; the point B will be in the parabola, as is evident.

Cor. 2. If the right line AB terminated by the parabola be ordinately applied to the diameter FL, and if the abscifs FG, between the ordinate and the vertex of the diameter, be equal to a fourth part of the parameter of the fame diameter FL; the line AB will be equal to the parameter. For by this Prop. AG, viz. half of AB is a mean proportional between the parameter and abscifs GF, viz. the fourth part of the parameter; therefore AG is the half of the parameter, and therefore AB is equal to the parameter.

PROP. II.

If a right line touching a parabola meets a diameter; the square of the segment between the point of contact and this diameter will be equal to the rectangle under the segment of this diameter, between its vertex and the tangent, and the parameter of the diameter which passes through the point of contact.

LET the right line FR touch the parabola in F, and meet the diameter ED in the point R, and draw the diameter FG through the point of contact, and let the right line P be its parameter; the square of FR will be equal to the rectangle under DR and the line P. Fig. 1.

For draw from the vertex of the diameter ED the line DL ordinately applied to the diameter FG; it will be parallel and equal to the tangent FR, and its abscifs LF will be equal to DR: then because (by the preced. Prop.) the square of DL is equal to the rectangle under the line LF and the line P; the square of FR will be equal to the rectangle under DR and the line P. *Q. E. D.*

PROP.

PROP. III.

If a right line cutting a parabola meets a diameter; the rectangle under its segments between the diameter and parabola will be equal to the rectangle under the segment of the diameter, between its vertex and the secant, and the parameter of the diameter to which the secant is ordinately applied.

Fig. 1. LET a right line cut a parabola in the points A, B, and meet the diameter ED in E, and let FG be the diameter to which AB is ordinately applied, and the right line P its parameter; the rectangle AEB will be equal to the rectangle under ED and the line P.

For through the vertex of the diameter FG draw a right line touching the parabola, and meeting the diameter in R, it will be parallel to the secant AB (by Cor. 1. 25. Book I.) and therefore (by Cor. 19. Book I.) the rectangle AEB is to the square of FR as ED to DR, or as the rectangle under ED and the line P to that under DR and the line P; therefore, by permutation, the rectangle AEB is to the rectangle under ED and the line P as the square of FR to the rectangle under DR and the line P; that is in a ratio of equality by the preced. Prop. *Q. E. D.*

Corollaries to the two preceding Propositions.

Cor. 1. If two right lines touching a parabola meet each other; the squares of their segments, between the points of contact and concourse, will be to each other as the parameters of the diameters which pass through their points of contact. For (a diameter being drawn through the point of concourse of the tangents) the square of either of the tangents will be equal to the rectangle under the parameter of the diameter which passes through its point of contact, and the same right line, viz. the segment of the diameter between its vertex and the point of concourse of the tangents.

Cor.

Cor. 2. And in like manner, if two right lines cutting a parabola meet each other; the rectangles under the segments of the secants between their point of intersection and the parabola, will be to each other as the parameters of the diameters to which the secants are ordinately applied, this appears from the preced. Prop.

Cor. 3. Or if a right line touching a parabola meets a secant, it is manifest from the two preceding Propositions, that the square of the segment of the tangent between the points of concourse and contact, will be to the rectangle under the segments of the secant between the point of concourse and the parabola as the parameter of the diameter which passes through the point of contact, to the parameter of the diameter to which the secant is ordinately applied.

Cor. 4. Because the rectangle AEB under the sum and difference of the ordinates BG, DL is equal to the rectangle under the parameter P and ED, the difference of the abscisses; the parameter of any diameter will be to the sum of the two ordinates, as their difference to the difference of the abscisses.

PROP. IV.

If from a point in an ellipse, an ordinate FG be applied to the diameter AB; the square of the ordinate will be equal to the rectangle under one abscis GB and a part of the parameter of the diameter AB, which is to the whole as the other abscis AG to the diameter AB. Fig. 2.

FROM the vertex B draw BH perpendicular to the diameter and equal to its parameter; let AH be joined, and draw from the point G a line parallel to BH, meeting AH in K, and let the rectangle BGKL be completed; the square of FG will be equal to this rectangle.

For by Cor. 1. to Def. II. the rectangle AGB is to the square of FG as the diameter AB to its parameter BH, or (because GK is parallel to BH) as AG to GK, that is as the same rectangle AGB to the rectangle KGB; therefore the square of FG is equal to the rectangle

KGB under the abſciſs GB and the line KG, which is to the parameter BH as the other abſciſs AG to the diameter AB.

Cor. Let the rectangles KLHM, ABHN be completed; it is manifeſt that the ſquare of the ordinate FG is deficient from the rectangle GH under the abſciſs GB and the parameter of the diameter AB, by a rectangle KH ſimilar and ſimilarly placed to the rectangle AH under the diameter AB and its parameter BH. From this defect Apollonius called this ſection an *Ellipſe*.

PROP. V.

Fig. 3. If an ordinate FG be drawn from a point in an hyperbola to the tranſverſe diameter AB; the ſquare of the ordinate will be equal to the rectangle under the leſſer abſciſs GB and a right line which is to the parameter of the diameter AB, as the other abſciſs AG to the diameter AB.

FROM the vertex B draw BH perpendicular to the diameter and equal to its parameter; join AH, and draw through the point G a line parallel to BH, meeting AH in K, and complete the rectangle KGBL; the ſquare of FG will be equal to this rectangle.

For by Cor. 1. to Def. II. the rectangle AGB is to the ſquare of FG as the diameter AB to its parameter BH, or (becauſe GK is parallel to BH) as AG to KG, that is as the ſame rectangle AGB to the rectangle KGB; therefore the ſquare of FG is equal to the rectangle KGB under the leſſer abſciſs GB and the line KG, which is to the parameter BH as the greater abſciſs AB to the diameter AB. *Q. E. D.*

Cor. Complete the rectangles KLHM, ABHN, and it is manifeſt that the ſquare of the ordinate FG exceeds the rectangle GH under the abſciſs GB and the parameter of the diameter AB, by a rectangle

rectangle similar and similarly placed to the rectangle AH, which is contained by the diameter AB and its parameter BH. From this excess, Apollonius called this section an *Hyperbola*.

PROP. VI. PROBL. I.

A parabola being given in position; to find its axis and to demonstrate there can be one axis only.

LET FBL be the parabola, aad find any diameter VO, and draw through a point X below the vertex a perpendicular to VO meeting the parabola in F, L; if the line FL be bisected in X, the diameter VO will be the axis, by Def. XXII. Book I. but if it be not bisected in X, draw another diameter AB bisecting FL in E, this diameter will be the axis; because it cuts the ordinate FL at right angles, for AB is parallel to VO. FIG. 4.

But if two diameters could be axes, the line terminated by the parabola and perpendicular to both of them, would be an ordinate to each of them, which is absurd: therefore there is one axis only to the same parabola. *Q. E. D.*

PROP. VII. PROBL. II.

Two conjugate diameters of an ellipse or hyperbola being given in position and magnitude; to find their axes, and to demonstrate that there can be two axes only in each section.

LET the right lines FL, GH, not at right angles to each other, be conjugate diameters of an ellipse or hyperbola, and C the center: through F the vertex of one of the diameters, which in an hyperbola is a transverse diameter, draw FV parallel to GH, and take from the same point F in the diameter FL the segment FK (towards the center in the hyperbola, and the contrary way in the ellipse) so that the rectangle CFK be equal to the square of CG, and KC being bisected in P, draw PQ perpendicular to KC, and meet- FIG. 5, 6.

ing the line FV in Q; then deſcribe a circle from the center Q paſſing through the points C, K, and meeting the line FV in the points D, O; the right lines CD, CO will be the two axes.

For by Cor. 27. Book I. the line FV touches the ſection, of which FL, GH are conjugate diameters, and by a property of the circle, the rectangle DFO is equal to the rectangle CFK, that is to the ſquare of the ſemidiameter CG (by conſtruction) therefore CD, CO are conjugate diameters (Cor. 51. Book I.) and cut each other at right angles, becauſe the angle DCO is in a ſemicircle, and therefore they are axes (Def. XXII. Book I.) for they cut their ordinates at right angles: the two vertices are thus found; draw from the point F the right line FE to the axis CD, and parallel to the other axis, and let CD, CB, CE be proportional, and take CA equal to CB; the points B, A will be the two vertices of the axis CD, (by 48. 49. Book I.) in like manner the two vertices M, N of the other axis are found. *Q. E. I.*

The above conſtruction remaining, if two other conjugate diameters as CX, CV could be axes, let them meet the tangent FD in the points X, V; the rectangle XFV will be equal to the ſquare of CG (51. Book I.) that is to the rectangle CFK; therefore the circle deſcribed through the points C, K, X will alſo paſs through V, and becauſe, by hypotheſis, XCV is a right angle, the center of this circle will be in the line FXV, and becauſe the line PQ biſects CK perpendicularly, the center of the circle will be likewiſe in the line PQ, and therefore in the point Q, that is in the center of the circle paſſing through the points C, K, D; therefore two circles, cutting each other in the points C, K, have the ſame center, which is abſurd: therefore there can be but two axes in each ſection. *Q. E. D.*

If in the hyperbola the diameters FL, GH were equal, the point K would coincide with the point C, and the circle deſcribed from the center Q would touch FL in C, and the demonſtration of this caſe is the ſame as the preceding, by only ſubſtituting the ſquare of FC inſtead of the rectangle CFK.

If the given conjugate diameters are perpendicular to each other, they are axes, as is evident.

Cor.

Cor. 1. Hence, an ellipse or hyperbola being given in position their axes may be found, for the center is found by Cor. 7. 25. Book I. and two conjugate diameters by the 27. Book I. and the magnitude of a second diameter of the hyperbola is determined by Def. XXIV. Book I. Or the section being given in position and its center being found, the axes may be more readily found: for if a circle whose center is the same as the center of the section, be described meeting the section in two points, and these points be joined by a right line, the diameter of the section bisecting this line will be an axis, as is evident.

Cor. 2. It is manifest from Cor. 8. 25. Book I. that a right line drawn through the vertex of an axis of a conic section and perpendicular to it, touches the section in that point; and on the contrary, if a right line touches the section at the vertex of the axis, it will be perpendicular to it.

PROP. VIII.

The axes of an ellipse are unequal; and the greater axis is the greatest, and the lesser the least of all the diameters of the ellipse; and the axes of an hyperbola are the least of all its diameters.

Part 1. LET AB, MN be the axes of an ellipse, they will be unequal by Cor. 2. 31. Book I. Let AB be the greater, which is called the transverse axis, and MN the lesser, which is called the second or conjugate axis: any semidiameter as CF is less than CB, but greater than CM; for let an ordinate FE be applied to AB; and because CEF is a right angle, the square of CF is equal to the squares of CE, FE together; but the square of FE is less than the rectangle AEB; for it is to that rectangle as the square of CM to the square of CB (Cor. 1. 31. Book I.) therefore the square of CF is less than the square of CE and the rectangle AEB together, that is, than the square of CB (5. 2.) therefore CF is less than CB; and in the same manner, an ordinate FL being FIG. 7.

 drawn

drawn to the axis MN, it may be demonſtrated that the ſemidiameter CF is greater than CM.

FIG. 6. *Part* 2. Let an ordinate FE be drawn from the vertex of the diameter CF of the hyperbola to the tranſverſe axis AB; and becauſe FEC is a right angle, the ſemiaxis CB is leſs than CF, and in like manner it may be ſhewn that the ſecond ſemiaxis CM is leſs than any other ſecond ſemidiameter; for CM is the tranſverſe axis of the hyperbola, conjugate to FB.

COR. 1. It is evident that the more remote any diameter is from the axis of the hyperbola, the greater it is.

FIG. 7. COR. 2. The diameter CF nearer to the greater axis of the ellipſe is greater than the diameter CP more remote. For if CF, CP were equal, the diameter biſecting the right line FP would cut it at right angles and would be an axis, contrary to what has been proved by the preceding; and if CF be leſs than CP, the circle deſcribed from the center C with the diſtance CF would cut CP within the ellipſe, but it meets the leſſer axis without the ellipſe, and therefore meets the ellipſe ſomewhere between the points F, M, ſuppoſe in Q; then CF, CQ would be equal, and the diameter biſecting FQ would be an axis, as before (contrary to the preced.) Hence the Corollary is evident.

COR. 3. Hence, it is manifeſt that two diameters of an ellipſe or hyperbola, equally diſtant from the ſame axis are equal; and converſely.

FIG. 8. COR. 4. The axes AB, MN of an hyperbola biſect the angles contained by the aſymptotes CY, CZ. For draw through the vertex of the axis AB a line parallel to the other axis, it will touch the hyperbola; let it meet the aſymptotes in the points R, S; RB, BS will be equal (36. Book I.) and ſince the angles at B are right angles, the triangles RBC, SBC will be equiangular, and therefore the angles at C are equal.

DEFI-

DEFINITIONS III, IV, V.

III. IF a point in the axis of a parabola be taken within the section, and its distance from the vertex of the axis be equal to a fourth part of its parameter; that point is called the *Focus* of the parabola.

IV. If two points F, O, in the transverse axis of an ellipse or hyperbola, be taken within the ellipse or opposite hyperbolas, so that the rectangle AFB or AOB be equal to the square of the conjugate semiaxis CM; each of these points F and O is called the *Focus* of the ellipse or opposite hyperbolas, and when the two points are mentioned together they are called the *Foci*. FIG. 9, 10.

V. Through the focus of a conic section let an ordinate be drawn to the axis, and through the point where it meets the section, let a tangent be drawn; this tangent is called the *Focal tangent.*

COR. 1. If from the vertex M of the conjugate axis of an ellipse a circle be described with an interval equal to the transverse semiaxes CA; it will cut the transverse axis in the foci: for let it meet it in the points F, O; the square of MF or CA is equal to the squares of CF and CM together: but the square of CA is equal to the square of CF and the rectangle AFB together (5. 2.) therefore the rectangle AFB is equal to the square of CM; and consequently F is a focus; and in like manner O will be the other focus. FIG. 9.

COR. 2. If from the center C, and in the transverse axis of an hyperbola, a segment CF or CO be taken equal to the right line AM joining two vertices of the axes; the extremity of this segment will be a focus. FIG. 10.

For the square of AM or CF is equal to the squares of CA, CM together, and (by 6. 2.) the same square of CF is equal to the square of CA and rectangle AFB; therefore the rectangle AFB is equal to the square of CM, and consequently F is a focus, and in like manner O will be the other focus.

COR. 3. It is evident that the foci of an ellipse, or of opposite or conjugate hyperbolas, are equally distant from the center.

PROP.

PROP. IX.

FIG. 9, 10, 11. Draw through the focus F of the conic section RAP, FT an ordinate to the axis AB, and draw the focal tangent TD; if a tangent be drawn through the vertex of the axis AB, the segment intercepted between the point of contact and the focal tangent, will be equal to the segment AF of the axis between the same point of contact and the focus F.

FIG. 11. FIRST, let the section be a parabola, and let the focal tangent cut the axis in D, and meet the tangent AH, drawn through the vertex of the axis, in the point H; AH, AF will be equal.

For (by Prop. XLVII. Book I.) AD is half of FD; therefore AH will be the half of FT; but FT is half of the parameter of the axis (Def. III. and Cor. 2. 1. of this Book) therefore AH is a fourth part of the same parameter, and consequently is equal to AF.

FIG. 9, 10. Secondly, let the section be an ellipse or hyperbola, and let the focal tangent meet the axis AB in D; and the tangents drawn through the vertices of the axis AB in H and G; AH, BG will be equal to AF, BF; for the rectangle under AH, BG will be equal to the square of the conjugate semiaxis CM (by 50. Book I.) that is to the rectangle AFB (by Def. IV.) and by (Cor. 5. 18. Book I.) AH is to BG as (HT to TG, or on account of parallel lines as) AF to FB; therefore these equal rectangles under AH, BG and AF, FB are also similar, and consequently their homologous sides AH, AF and BG, BF are equal. *Q. E. D.*

COR. The above construction remaining, if a right line be drawn from any point P of a conic section or opposite section to the focus F, and the perpendicular PQ to the axis AB, meeting the focal tangent in L, and the section again in R; the square of FQ will be equal to the rectangle PLR; for because the tangent AH is parallel to the secant LPR, the square of AH will be to the rectangle PLR as (the square of HT to the square of TL (Cor. 6. 18.

Book

Book I.) or on account of parallel lines) as the square of AF to the square of FQ: but the square of AH is equal to the square of AF by this Prop. therefore the rectangle PLR is equal to the square of FQ.

PROP. X.

Through the focus F of the conic section RAP, draw FT an ordinate to the axis AB, and draw the focal tangent TD; and a right line from any point P in this section or in the opposite section, to the focus F, and the perpendicular PQ to the axis AB, meeting the focal tangent TD in L; PF, LQ will be equal. FIG. 9, 10, 11.

LET the line PQ meet the section again in R; then because FQP is a right angle, the square of FP is equal to the squares of PQ, FQ, that is (by Cor. preced.) to the square of PQ together with the rectangle PLR, that is to the square of LQ (6. 2.) therefore the line FP is equal to LQ. *Q. E. D.*

DEFINITION VI.

LET the focal tangent TH meet AB the axis of a parabola, or the transverse axis of an ellipse or hyperbola in the point D, and draw through the point D a perpendicular to AB: this right line is called the *Directrix* of the conic section. FIG. 9, 10, 11.

COR. 1. Every right line perpendicular to the directrix of the parabola is a diameter, and conversely, as is evident from Cor. 5. to the Definitions of the sections, Book I.

COR. 2. The distance of the directrix from the vertex of the parabola is equal to the distance of the focus from the same vertex, by Prop. XLVII. Book I.

COR. 3. The semiaxis CA in the ellipse or hyperbola is a mean proportional between the distance CD of the directrix from the center,

center, and the distance CF of the focus from the center, as is evident from Prop. XLVIII. and XLIX. Book I. It is likewise evident that there are two directrices in the ellipse and hyperbola equally distant from the center.

PROP. XI.

FIG. 9, 10, 11. **If from any point P in a conic section RAP, a right line PE be drawn perpendicular to the directrix DE, and the right line PF to the focus nearer to the directrix; PF drawn to the focus will be to the perpendicular PE as FA, the distance of the focus F from the nearer vertex of the axis, to DA, the distance of the directrix DE from the same vertex.**

FOR draw from the point P a right line PQ perpendicular to the axis AB, and meeting the focal tangent DT in L; the segment QD of the axis between PQ and the directrix will be equal to PE, and LQ is equal to PF (by Prop. preced.) draw through the vertex A of the axis a tangent meeting DT in H; then because AH, LQ are parallel, LQ will be to QD as HA to AD, that is as FA to AD (9. of this Book) therefore PF is to PE as FA to AD. *Q. E. D.*

COR. 1. Hence, in the ellipse and hyperbola, the right line PF drawn to the focus is to PE perpendicular to the directrix as CF, the distance of the focus from the center, to CA the transverse semiaxis. For by Cor. 3. preced. CF is to CA as CA to CD; therefore by conversion CF is to FA as CA to AD, and, by permutation, CF is to CA as FA to AD, that is as PF to PE by this Prop. therefore PF is to PE as CF to CA.

COR. 2. From the preceding demonstration, the distance CF of the focus from the center is to the semiaxis CA as the distance FA of the focus from the nearer vertex to the distance DA of the directrix from the same vertex.

COR.

Cor. 3. Right lines drawn from points in the same section or opposite sections to a focus, are to each other as the perpendiculars drawn from the same points to the directrix nearer to that focus.

SCHOLIUM.

LET DAH be a right angled triangle, and take in the base DA produced beyond the right angle A, a line FA equal to the perpendicular HA, and let the line FA revolve about the point F, and increase in such a manner, that it be always equal to the line drawn through its extremity parallel to HA, and terminated by the lines DA, DH produced: the line described by the motion of the extremity of this revolving line will be a conic section, viz. an hyperbola, a parabola or an ellipse, according as the base DA of this triangle may be less, equal, or greater than the perpendicular HA; hence the point D, the intersection of the base with the hypothenuse, receding ad infinitum, *the hypothenuse will at last be parallel to the base, and the ellipse will be changed into a circle whose center is F; in the other cases, the point F will be a focus of the section described, and the point A the vertex of its axis; and when the revolving line shall be parallel to HA, its extremity will be in the line DH, which will therefore touch the section. All which is manifest from what has been demonstrated above: from hence likewise the point B, the other extremity of the transverse axis, and center C may be found, when the section is an ellipse or hyperbola.* Fig. 9, 10, 11.

PROP. XII.

A right line drawn from any point P in a parabola to the focus F, is equal to the perpendicular PE drawn from the same point to the directrix. Fig. 11.

FOR, by the preceding, PF is to PE as FA to AD, that is, in a ratio of equality, by Cor. 2. Def. VI. of this Book. *Q. E. D.*

Cor. 1. The distance of a point within a parabola from the focus is less, and of a point without a parabola is greater, than the perpendicular drawn from the same point to the directrix. Let the point M be within the parabola, and draw MF to the focus, and from

 the

the ſame point a perpendicular to the directrix meeting it in E, and the parabola in P, and join PF; becauſe MF is leſs than MP, PF together, it will be leſs than MP, PE, that is, than ME. Now let the point N be (without the parabola) in the line PE perpendicular to the directrix, and join NF; NF and NP together are greater than PF or PE; therefore NF is greater than NE.

Cor. 2. Hence, if the diſtance of any point from the focus of a parabola be equal, leſs, or greater than the perpendicular drawn from the ſame point to the directrix, that point, in the firſt caſe, will be in the parabola, in the ſecond caſe, within the parabola, and in the third caſe, without it.

PROP. XIII.

Fig. 10. **If from a point P in the hyperbola RAP a right line PF be drawn to the focus F, and the right line PN, parallel to the aſymptote CY, be drawn to the directrix DE, nearer to the focus F; theſe lines PF, PN will be equal.**

FOR let the perpendicular PE be drawn from the point P to the directrix DE, and let O be the other focus, and draw through the vertex B of the tranſverſe axis a tangent meeting the aſymptote CY in the point Y, BY will be equal to the conjugate ſemiaxis CM (XXXVIII. Book I.) and CY will be equal to CO the diſtance of the focus from the center (by Cor. 2. Def. IV.) and (by Cor. 1. XI. of this Book) PF is to PE as (CO, or) CY to CB: but becauſe the triangles PEN, CBY are ſimilar, PN is to PE as CY to CB; therefore PF, PN are equal. *Q. E. D.*

Cor. It may be ſhewn in the ſame manner as in the preceding Cor. that the diſtance of a point, within an hyperbola, from the focus, is leſs, and from a point taken without the hyperbola, greater than the line drawn parallel to the aſymptote from the ſame point to the directrix nearer to this focus; and from hence, if the diſtance of any point from the focus of an hyperbola be equal to the line drawn from the ſame point to the directrix nearer to this focus and parallel to the aſymptote; that point will be in the hyperbola, to the focus of which the right line was drawn.

PROP.

PROP. XIV.

If two right lines be drawn from a point in an ellipse or hyperbola to the foci; the sum of these lines in the ellipse, and their difference in the hyperbola, is equal to the transverse axis.

LET AB be the transverse axis of an ellipse or hyperbola, and the points F, O the foci, and DE and ZX the directrices; draw from a point P in the section PF and PO to the foci, and a line parallel to AB, meeting the directrices DE, ZX in the points E and X. Fig. 9, 10.

(By Cor. 1. Prop. XI. of this Book) PF is to PE as CF to CA, that is, as CA to CD (Cor. 3. Def. VI.) or as AB to DZ; and in like manner, PO is to PX, as AB to DZ; therefore the sum or difference of the antecedents PF, PO is to the sum or difference of the consequents PE, PX as AB to DZ; but in the ellipse the sum and in the hyperbola the difference of the consequents PE, PX is equal to DZ; therefore in the ellipse the sum and in the hyperbola the difference of the antecedents PF, PO is equal to the transverse axis AB. *Q. E. D.*

Cor. 1. If two right lines be drawn from a point without the ellipse to the foci; these two lines together will be greater than the transverse axis: but if they be drawn from a point within the ellipse they will together be less than the same axis, as is evident from the 20, 1.

Cor. 2. And on the contrary, if the lines drawn from any point to the foci, are together equal, greater, or less than the transverse axis; that point in the first case will be in the ellipse, in the second case without it, and in the third case within the ellipse.

Cor. 3. If right lines MF, MO be drawn from a point M within the hyperbola LAP to the foci; their excess will be greater than the transverse axis AB: but if they be drawn from a point N without the hyperbola, their excess will be less than the same axis: for let the Fig. 11.

the point M be within the hyperbola, whose focus is F, and let the line MO drawn to the other focus O, meet this hyperbola in L, and join LF; the excess of the line MO above MF is greater than its excess above LM, LF together, that is, than the excess of the line LO above LF, that is, than the transverse axis by this Prop.

Now let the point N be without the hyperbola, and draw from the point L, where the line NF meets the hyperbola, a right line to the focus O; NO will be less than NL and LO together; therefore the excess of NO above NF is less than the excess of NL, LO above NF, that is, than the excess of the line LO above LF, that is, than the transverse axis AB.

Cor. 4. And on the contrary, if the excess of the lines, which are drawn from any point to the foci of the hyperbolas, be equal, greater, or less than the transverse axis; this point will be in one of the hyperbolas, or within, or without the hyperbola.

SCHOLIUM.

WHILE I endeavoured to deduce from the nature of the cone, the origin of those lines that are called the directrices of the sections; a theorem occurred to me, which seemed not an inelegant one, and in some measure answered my design. But as it would here interrupt the series of the Propositions, it is inserted at the end of this Book.

PROP. XV.

Fig. 13. If a right line PF be drawn from any point P in a parabola to the focus, and a perpendicular PE to the directrix; the right line PT bisecting the angle FPE contained by these lines will touch the parabola: and on the contrary, the line PT touching the parabola in P will bisect the angle contained by the lines PF, PE.

Part 1. LET any point X be taken in the line PT, and draw the right lines XF, XE and EF, and draw the perpendicular XD to the directrix; then because in the triangles PTF, PTE,

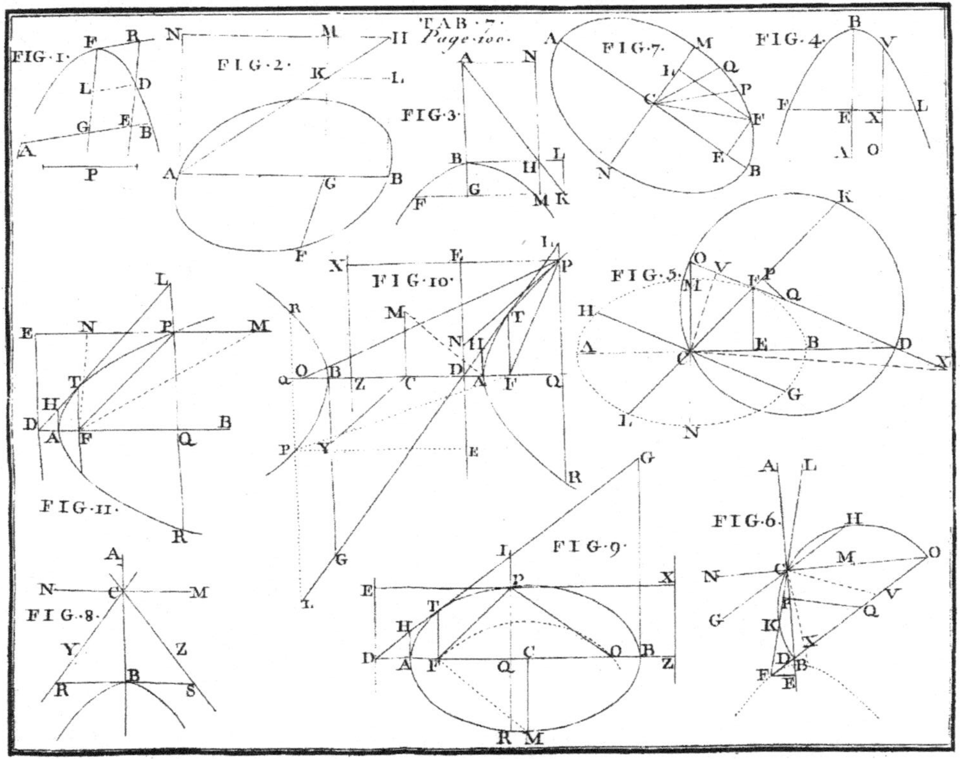
TAB·7·
Page 100·
FIG·1·
FIG·2·
FIG·3·
FIG·4·
FIG·5·
FIG·6·
FIG·7·
FIG·8·
FIG·9·
FIG·10·
FIG·11·

PTE, PF, PE are equal (XII. of this Book) and PT common to both, and the angles TPF, TPE equal, TF, TE will be equal, and the angles at T right angles; therefore XF is equal to XE, and consequently greater than XD (19. 1.) therefore the point X is without the parabola (by Cor. 2. XII. of this Book) and therefore the line PT touches the parabola in P.

Part 2. If now the line PT touches the parabola in P, it will bisect the angle FPE. For if not, draw another line bisecting this angle, this line likewise will touch the parabola in P, by the first case, which is impossible (XVI. Book I.) therefore the tangent PT bisects the angle FPE. *Q. E. D.*

PROP. XVI.

If two right lines PF, PO be drawn from any point P in an ellipse to the foci; the right line PT bisecting the angle FPE, which is adjacent to the angle FPO contained by the lines drawn to the foci, will touch the ellipse in the point P: and on the contrary, if the line PT touches the ellipse in P, it will bisect the angle FPE adjacent to the angle FPO contained by the lines drawn from the point P to the two foci. FIG. 14.

Part 1. LET OP be produced to E, so that PE be equal to PF, and let a point X be taken in the line PT, and draw XO, XF, XE, and also FE meeting PT in T; then because in the triangles PTF, PTE, PF, PE are equal, and PT common to both, and the angles TPF, TPE equal, TF, TE will be equal, and the angles at T right angles; therefore XF, XE will be equal; but XO, XE together are greater than OE; therefore XO, XF together will be greater than OE, that is, than OP, PF, that is, than the transverse axis of the ellipse (XIV. of this Book) therefore the point X is without the ellipse (by Cor. 2. XIV. of this Book) and consequently the line PF touches the ellipse in the point P. *Q. E. D.*

Part 2. May be demonstrated in the same words as the second part of the preced. Prop.

PROP.

PROP. XVII.

FIG. 12.

If two right lines PF, PO be drawn from a point P in an hyperbola to the foci, the right line PT bisecting the angle FPO contained by these lines, touches the hyperbola in P: and on the contrary, if the line PT touches the hyperbola in P, it will bisect the angle FPO contained by the same lines drawn from the point P to the foci.

Part 1. IN the greater right line PO take PE equal to PF; EO will be equal to the transverse axis (XIV. of this Book) join FE meeting PT in T, and take any point X in the line PT, and draw XO, XF, XE, and as in the preceding, XE, XF will be equal: and because XO is less than XE, EO together, its excess above XF will be less than the excess of XE, EO above XF, that is than EO, which is equal to the transverse axis; and consequently the point X is without the hyperbola (Cor. 4. XIV. of this Book) and therefore the line PT touches the hyperbola in the point P.

Part 2. Is demonstrated in the same words as the second part of Prop. XV. Q. E. D.

SCHOLIUM.

THE foci of the ellipse and hyperbola were called by Apollonius, Puncta ex comparatione facta.

The focus of the parabola he does not mention. Modern writers have named these points Umbilici or Foci. Perhaps from hence; that the rays of light falling on a concave speculum, formed by the revolution of a conic section round its principal axis, will all meet after reflection in the focus of the section, whose revolution generates the speculum.

For if the rays fall on a parabolic speculum and are parallel to its axis, they will meet in its focus after reflection: Or if they fall on an elliptic speculum, and are diverging from one focus, they will meet in the other focus after reflection; and if they fall on an hyperbolic speculum, and are converging towards the focus of the hyperbola opposite to the one which gene-

generates the speculum, they will meet in its focus after reflection. All which appears from the three foregoing Propositions, and from hence that the angle of reflection is equal to the angle of incidence.

PROP. XVIII.

Draw through the extremities of the transverse axis AB of an ellipse or hyperbola two tangents which may meet any other tangent PT in the points H, G; a circle described about the diameter GH will pass through the foci F, O of the section. FIG. 15, 16.

LET CM be the conjugate semiaxis, and because the tangents AH, BG are parallel, and meet a third tangent PT in the points H, G; the rectangle under AH, BG will be equal to the square of CM (by L. Book I.) that is, to the rectangle AFB (by Def. IV. of this Book) therefore AH is to AF as FB to BG, and these lines are about the equal angles FAH, FBG; therefore the triangles FAH, FBG are equiangular (*6. 6.*) and therefore the angles FHA, BFG are equal; but the angles FHA, HFA are equal to a right angle (for FAH is a right angle); therefore the angles BFG, HFA are equal to a right angle; and therefore GFH is a right angle, and consequently the point F is in the periphery of the circle described about GH (converse. 31. 3.) in like manner it may be shewn that the other focus O is in the periphery of the same circle. *Q. E. D.*

PROP. XIX.

If two right lines PF, PO be drawn from a point P in an ellipse or hyperbola to the foci; they will contain a rectangle equal to the square of the semidiameter CD parallel to the tangent drawn through the point P. FIG. 15, 16.

FOR draw through the extremities of the transverse axis the tangents AH, BG meeting the tangent drawn through P in the points H, G: by the preceding, a circle described about the diameter

meter GH will pass through the foci; draw from either focus as F, the perpendicular FT to GH, and let it meet OP in E: and because the angles FPT, EPT are equal (by Part 2. Prop. XVI. XVII. of this Book) and PT common; FT will be equal to TE and PF to PE; therefore because the diameter GH of the circle bisects FE and is perpendicular to it, and the point F is in its circumference, the point E will be in the same; and consequently by a property of the circle, the rectangle OPE, that is, OPF is equal to the rectangle GPH, that is, to the square of the semidiameter CD parallel to GH (by Part 2. Prop. L. Book I.) *Q. E. D.*

PROP. XX.

Fig. 15, 16. **If from the foci F, O of an ellipse or hyperbola, the right lines FT, OL be drawn perpendicular to the tangent GPH; a circle described about the transverse axis will pass through the points T, L, where the perpendiculars meet the tangent.**

LET AB be the transverse axis and C the center, and draw PF, PO from the point of contact P to the foci, and let the perpendicular FT meet PO in E, and join CT. Because the angles FPT, EPT are equal (XVI. and XVII. of this Book) and PT common, FP will be equal to PE; therefore OE is equal to the transverse axis (XIV. of this Book) but FT, TE are equal, and likewise FC, CO (Cor. 3. Def. V. of this Book) therefore OE, CT are parallel, and consequently because CF is half of OF, CT will be the half of OE, that is, half the transverse axis AB: therefore the point T is in the circumference of the circle described about AB: and in like manner the point L will be in the same circumference. *Q. E. D.*

PROP.

PROP. XXI.

If the perpendiculars FT, OL be drawn from the two foci F, O to any tangent GPH; they will contain a rectangle equal to the square of the conjugate semiaxis. Fig. 15, 16.

FOR let AB be the transverse axis, and C the center, and join CT, which may meet OL in N; because ON, FT are parallel, the triangles CFT, CON will be equiangular, and therefore because CF, CO are equal, CT, CN, and likewise ON, FT will be equal: but the circle described about the axis AB will pass through the points F and L (by the preceding) and because CT, CN are equal, it will likewise pass through the point N; and therefore by a property of the circle, the rectangle under LO (ON, or) TF will be equal to the rectangle BOA, that is, to the square of the conjugate semiaxis (by Def. IV. of this Book). *Q. E. D.*

PROP. XXII.

Let the right line PT touch an ellipse or hyperbola, and at the point of contact P draw PN, meeting the transverse axis, and perpendicular to the tangent, and draw from the center, CK perpendicular to the same tangent; these perpendiculars will contain a rectangle equal to the square of the conjugate semiaxis CM. Fig. 17, 18.

LET AB, M*m* be the axes of the ellipse or hyperbola, and C the center. Draw from the point of contact P the right lines PE, PQ perpendicular to the axes; QC will be equal to PE: let CM meet PT in R, and on account of parallel lines, the angles NPE, RCK will be equal; therefore the right angled triangles CKR, PEN are similar; and CK will be to CR as PE to PN, and consequently the rectangle under CK and PN is equal to the rectangle under CR and PE or CQ, that is, to the square of CM (by XLVIII. XLIX. Book I). *Q. E. D.*

PROP. XXIII.

Fig. 15, 16. Let PT touch an ellipse or hyperbola, and draw from the point of contact P, the right line PO to the focus, and from the center, CT parallel to PO, meeting the tangent; CT will be equal to half of the transverse axis AB.

FOR draw PF to the other focus and join FT, and let it be produced that it may meet PO in E: then because OE, CT are parallel and OC, CT equal; ET, TF will be equal; and because the tangent PT bisects the angle FPE, FP will be to PE as FT to TE (by 3. 6.) therefore FP, PE are equal, and therefore the line OE is equal to AB (by XIV. of this Book) but because CF is half of OF, CT will be the half of OE, or of the transverse axis AB. *Q. E. D.*

PROP. XXIV.

Fig. 17, 18, 19. Let the right line PT touch an ellipse or hyperbola, and draw from the point of contact P, two right lines to the axis of the section, one of them PE, an ordinate to the axis, and the other PN perpendicular to the tangent; the segment CE of the axis between the center and ordinate will be to the segment NE of the same axis, between the ordinate and perpendicular to the tangent, as the axis to its parameter.

LET the tangent PT meet the axis of the section AB in D, and first let AB be either of the axes of the ellipse or the transverse axis of the hyperbola; CE is to NE as the rectangle CED to the rectangle NED; but the rectangle CED is equal to the

the rectangle BEA (Cor. 1. XLIX. Book I.) and becaufe NPD is a right angle, and PE perpendicular to AB, the rectangle NED is equal to the fquare of PE; therefore CE is to NE, as the rectangle BEA to the fquare of PE, that is, as the axis AB to its parameter (by Cor. 1. Def. 2. of this Book).

Secondly, let AB be the conjugate axis of the hyperbola (by Cor. 2. XLIX. Book I.) the rectangle CED will be equal to the fquares of CE and CB, and, as before, the rectangle NED will be equal to the fquare of PE, and CE will be to NE, as the rectangle CED to the rectangle NED, that is, as the fquares of CE, CB to the fquare of the ordinate PE, that is, as the axis AB to its parameter (by Cor. II. Def. II. of this Book). FIG. 19.

PROP. XXV.

Let PT touch a parabola, and meet the axis AB in T, and draw from the point of contact P two right lines to the axis, one of them PE ordinately applied to the axis, and the other PN perpendicular to the tangent; the fegment EN of the axis intercepted by thefe lines will be half the parameter of the axis: and the fegment TN between the tangent and the perpendicular PN will be half the parameter of the diameter which paffes through the point of contact P. FIG. 20.

FOR draw through the point of contact P the diameter DQ, and by Prop. XLVII. Book I. the right line EA or TA is the half of TE.

Part 1. Becaufe NPT is a right angle and PE is perpendicular to TN, the rectangle under TE, EN is equal to the fquare of PE, that is, to the rectangle under AE and the parameter of the axis (I. of this Book) therefore as TE is to AE, fo is the parameter of the axis to EN, which is therefore the half of that parameter. This fegment EN of the axis is called the *Subnormal.*

Part 2. Because of the similar rectangled triangles TPN, TEP, the rectangle under TE, TN, will be equal to the square of TP, that is, to the rectangle under TA and the parameter of the diameter (Prop. II. of this Book) therefore TE is to TA as the parameter of the diameter DQ to TN, which is therefore the half of that parameter. *Q. E. D.*

Cor. 1. Hence, if a right line PF be drawn from the vertex P of any diameter DQ to the focus, it will be a fourth part of the parameter of that diameter. For, the tangent PT being drawn, because the diameter DQ is perpendicular to the directrix of the parabola, the angle FPT will be equal to the angle DPT (XV. of this Book) that is, to the alternate angle FTP; therefore FP, FT are equal: then if a circle be described from the center F through the points T, P, it will pass through N, because TPN is a right angle; therefore the line PF is half of TN; and therefore a fourth part of the parameter of the diameter DQ, by the second part of this Prop.

See Newton's Principia Math. Lemma 13. Book I.

Cor. 2. From hence the segment of any diameter between its vertex and directrix is a fourth part of its parameter, as is evident from (Cor. preced. and Prop. XII. of this Book) but if the point P were the vertex of the axis, both the Corollaries are evident from the Def. of the focus.

Cor. 3. If the right line joining the vertices of two diameters be an ordinate to the axis, their parameters will be equal. For their segments between the vertices and directrix are equal, being the opposite sides of a parallelogram.

PROP. XXVI.

A right line terminated by a conic section passing through the focus, and ordinately applied to the axis, is equal to the parameter of the axis: but every right line terminated by a parabola and passing through the focus, is equal to the parameter of that diameter to which the line is ordinately applied.

Part 1. FIRST let the section be an ellipse or hyperbola, whose transverse axis is AB and conjugate axis M*m*, the center C, and focus F, and through F let XY pass ordinately applied to the transverse axis AB; XY will be equal to the parameter of the axis AB. FIG. 17, 18.

For (by Cor. 1. Prop. XXXI. and Def. XXIV. Book I.) the square of CA is to the square of CM as the rectangle BFA, that is, the square of CM (Def. IV.) to the square of FY; therefore CA, CM, FY are proportional; therefore the whole lines AB, M*m*, XY are proportional, and therefore XY is the parameter of the transverse axis AB, by the definition of a parameter; secondly, if the section be a parabola, the ordinate to the axis passing through the focus is equal to the parameter of the axis, by Cor. 2. Prop. I. of this Book, and the Def. of the focus.

Part 2. Let XY terminated by a parabola pass through the focus and be ordinately applied to the diameter DQ and meet it in the point O; through the vertex P of this diameter draw PF to the focus, and let the line PT touch the parabola in P, it will be parallel to the ordinate OX; therefore the angle POF is equal to the angle DPT, or TPF (XV. of this Book) that is, to the alternate angle PFO; therefore the abscifs PO is equal to PF, that is, to a fourth part of the parameter of the diameter DQ, by Cor. 1. preceding; therefore XY is equal to the parameter of the diameter DQ, to which it is ordinately applied, by Cor. 2. Prop. I. of this Book. FIG. 20.

COR.

Fig. 18. Cor. Hence, if from the focus F of an hyperbola a right line FL be drawn parallel to an asymptote, till it meets the section, it will be a fourth part of the parameter to the transverse axis.

Draw through the focus F an ordinate FX applied to the axis, and meeting the hyperbola in X, and let *dr* be the directrix of the hyperbola, let FL produced meet the directrix in *d*, and draw X*r* to the directrix and parallel to F*d*; X*r* will be equal to F*d* (34. 1.) and by Prop. XIII. of this Book, FL, L*d* will be equal, as also FX, X*r*; therefore F*d* is equal to FX, that is, to half the parameter of the transverse axis (by Part 1. of this Prop.) and therefore FL the half of F*d* is a fourth part of the parameter of the transverse axis of the hyperbola.

PROP. XXVII.

Fig. 17, 18, 20. If a right line PT touches a conic section, and a right line PF be drawn from the point of contact P to the focus F, also PN perpendicular to the tangent, and meeting in N the axis AB, which passes through the focus, and if the perpendicular NV be drawn from the point N to PF; this line cuts from PF the segment VP equal to half the parameter of the axis AB.

Fig. 20. When the section is a parabola, draw PE perpendicular to the axis, because it has been shewn in Cor. 1. XXV. of this Book, that PF is equal to NF, the angle FPN is equal to the angle FNP; wherefore the right angled triangles PEN, NVP are equiangular, and have the side PN common, therefore these triangles are equal, and therefore PV is equal to NE, that is, to half the parameter of the axis (Part 1. Prop. XXV. of this Book).

Fig. 17, 18. But if the section be an ellipse or hyperbola, draw CK from the center, perpendicular to the tangent, and CT parallel to PF, and let CM be the conjugate semiaxis: and because CK, CT are parallel to PN, PV, the angles KCT, NPV will be equal; and therefore the right angled triangles CKT, NVP are equiangular; therefore the rectangle under CT, PV is equal to the rectangle under

CK

CK and NP, that is, to the ſquare of CM (XXII. of this Book) and therefore CT, CM, PV are proportional: but CT is equal to CA (by XXIII. of this Book) and conſequently PV is half the parameter of the axis AB, by the Def. of a parameter. *Q. E. D.*

PROP. XXVIII.

If through the focus O of an ellipſe or hyperbola a right line be drawn meeting the ſection or oppoſite hyperbolas in the points H, I; the diameter UZ drawn parallel to this line will be a mean proportional between the ſegment HI of the line intercepted by the ſection or hyperbolas, and the tranſverſe axis AB. Fig. 17, 18, 21.

LET C be the center of the ſection, and draw a diameter biſecting HI in G; it will be conjugate to UZ (by XXVII. Book I.) through the point H draw HS an ordinate applied to the diameter UZ, it will be parallel to CG (Cor. XXVII. Book I.) and draw the tangent HT meeting UZ in T: then becauſe CT drawn from the center to the tangent is parallel to HO joining the point of contact and focus; CT will be equal to CA: but CS is equal to GH (34. 1.) and the ſemidiameter CZ is a mean proportional between CT, CS (XLVIII. XLIX. Book I.) that is, between CA, GH; therefore the whole diameter UZ is a mean proportional between AB, HI. *Q. E. D.*

Cor. Hence, right lines terminated both ways by an ellipſe or hyperbola (and when neceſſary produced) paſſing through the focus, are to each other as the ſquares of the diameters to which they are parallel.

PROP.

PROP. XXIX.

If right lines terminated both ways by a conic section or opposite sections pass (produced when necessary) through the focus; the rectangles under their segments, viz. between the focus and section or sections, will be to each other as these right lines.

FOR if the section be an ellipse or hyperbola, the rectangles under the segments of these lines will be to each other as the squares of the diameters to which the lines are parallel (XXXI. and XL. Book I.) that is, as these lines by Cor. preced.

If the section be a parabola, the rectangles under the segments of the lines are as the parameters of the diameters to which these lines are ordinately applied (Cor. 2. Prop. III. of this Book) that is, as the lines themselves, by XXVI. of this Book. *Q. E. D.*

PROP. XXX.

If from the focus of a parabola a right line be drawn perpendicular to any tangent; the square of the perpendicular will be equal to the rectangle under the distance of the point of contact from the focus, and the distance of the focus from the vertex of the axis.

FIG. 13. LET AB be the axis of the parabola and F the focus, and draw through the point P in the parabola a tangent meeting the axis in N, draw the right lines PF to the focus, PB an ordinate to the axis, and PE perpendicular to the directrix: let FT be drawn from the focus perpendicular to the tangent, and join AT; because the angle NPF is equal to the angle NPE (by XV. of this Book) that is, to the alternate angle PNF, the lines FP, FN will be equal; therefore the perpendicular FT bisects PN in T, the segment BN is also bisected in the vertex A (by the XLVII. Book I.) therefore TA is is parallel to PB, and consequently is perpendicular to the axis, and therefore, because FTN is a right angle, the square of FT is equal to the rectangle under FA, FN or FP. *Q. E. D.*

Cor. Hence, the squares of perpendiculars drawn from the focus of a parabola to the tangents, are to each other as the distances of their points of contact from the focus.

PROP. XXXI.

If from the focus O of an ellipse or hyperbola, a right line OL be drawn perpendicular to any tangent GPH, and OP be drawn to the point of contact; OL will be to OP, as the conjugate semiaxis CM to the semidiameter CD, which is parallel to the tangent. Fig. 15, 16.

FROM the other focus F, draw FT perpendicular to the tangent GP, and PF to the point of contact; because (from the second parts of Prop. XVI. and XVII. of this Book) the angles FPT, OPL are equal, and consequently the right angled triangles FTP, OLP similar, and therefore the rectangles under OL, FT and OP, FP are similar; therefore the square of OL is to the square of OP, as the rectangle under OL, FT to the rectangle under OP, FP, that is, as the square of CM to the square of CD (by XXI. and XIX. of this Book) therefore OL is to OP, as CM to CD. *Q. E. D.*

Cor. 1. Hence, if from the center C of an ellipse or hyperbola, CK be drawn perpendicular to any tangent GH; CK will be to the transverse semiaxis CA, as the conjugate semiaxis CM to the semidiameter CD parallel to the tangent.

For draw from the focus the right line OL perpendicular to GH, and OP to the point of contact, and CT from the center to the tangent, parallel to OP; CT will be equal to CA (XXIII. of this

 Book)

Book) and because the triangles KCT, LOP are similar, CK is to CT or CA, as OL to OP, that is, as CM to CD, by this Prop.

Cor. 2. Because in the ellipse or hyperbola, the square of FT is to the square of FP, as the square of CM to the square of CD, and the square of the conjugate semiaxis CM is given; the square of FT will be as $\frac{FPq}{CDq}$ that is (by XIX. of this Book) as $\frac{FPq}{FP \times OP}$ or as $\frac{FP}{OP}$ therefore the square of FT increases in the ratio in which FP is increased, and OP diminished; but in the ellipse, when FP is increased, OP is diminished, and the contrary: therefore in the ellipse the square FT varies more than in the ratio of FP; but in the hyperbola FP, OP are increased together, or diminished together; therefore in the hyperbola the square of FT varies less than in the ratio of FP; and consequently the perpendicular FT drawn from the focus to the tangent in the ellipse varies more, and in the hyperbola less than in the subduplicate ratio of FP, viz. the distance of the focus F from the point of contact P.

See Newt. Prin. Math. Cor. 6. Prop. XVI. Book I.

PROP. XXXII.

Fig. 22. **Let AB be the transverse, and MN the conjugate axis of an ellipse, and C the center: if from a point D in the conjugate axis a right line DG be placed, meeting the transverse axis in G, and equal to the sum or difference of the semiaxes CA, CM, and if in DG (produced beyond the point G, when DG is the difference of the semiaxes) GP be taken equal to CM; the point P will be in the ellipse.**

FOR draw PE through the point P ordinately applied to the axis AB, and from the center draw a right line parallel to DP, and meeting PE produced in F; because CD, PF are parallel, CF is

TAB. 8. Page 114.

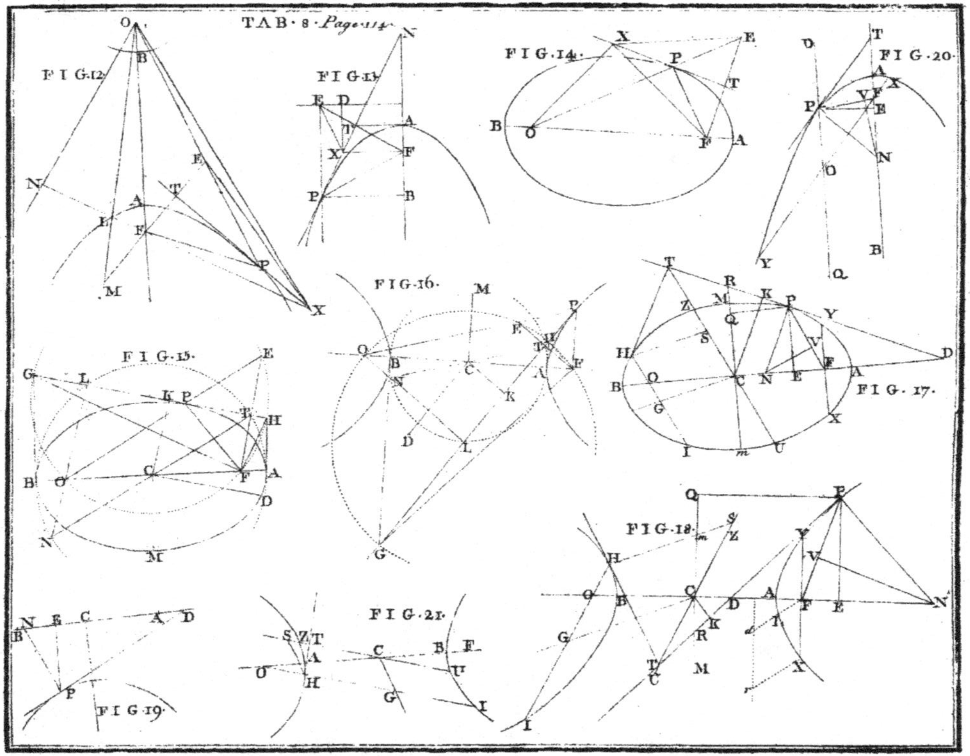

is equal to DP (34. 1.) that is, to CA by construction; therefore the point F is in the circumference of a circle described about AB; and because the triangles PEG, FEC are similar, the square of PE is to (the square of FE, or) the rectangle BEA, as the square of (GP or) CM to the square of (CF or) CA; therefore the point P is in the ellipse (by Cor. XXXIV. Book I.) *Q. E. D.*

Cor. Hence, if two right lines AB, MN bisect each other at right angles in the point C, and DG be equal to the difference of CA, CM, and GP (produced beyond the point G, which is in the greater line AB) be equal to CM, and the line DG be moved through four right angles so that the point D be always in the line MN and G in AB; the point P will describe an ellipse, whose transverse axis will be AB, and conjugate axis MN, and from hence an ellipse may be described by an instrument called the *Elliptical Compass*, as will readily appear by considering the construction of that instrument.

PROP. XXXIII. PROBL. III.

An indefinite right line being given in position, and a point given in it, and a given right line bisected by it; to describe a parabola of which this indefinite right line shall be a diameter, and the given point its vertex, and to which the given bisected line shall be an ordinate.

FIRST, let AB be the indefinite line given in position, bisecting the given line OS in C at right angles, and let A be the given point in the line AB. Fig. 13.

Find the line QR a third proportional to CA, CO, and from the point A in the line AB and towards the point C take a segment AF equal to the fourth part of QR, and let AD in the line AB be equal to AF, and draw through the point D the indefinite right line DX perpendicular to AB: then place the ruler HEG (whose sides HE and EG are perpendicular to each other) on the plane XDF, so that the side HE may be applied to DX, and let the other side EG be on the same side of DX with the point F, and let the extremity of a string GPF of the same length with the side EG be annexed to the end G of the side EG, and the other extremity of it fixt in the point F: let part of the string GP, by means of the pin P, be applied to the side of the ruler EG and stretched along it, then let the side HE of the ruler be moved along DX, and at the same time let the string, stretched by means of the pin, be constantly applied to EG the side of the ruler; the line described by the motion of the pin P will be the parabola required.

For let a parabola pass through the points O, A, S (by XXXIII. Book I.) of which let AB be a diameter and A its vertex, and to which OS be ordinately applied; AB will be the axis (by Def. XXII. Book I.) and the right line AF or AD taken four times, its parameter, by construction, and Def. I. of this Book, and consequently the point F is the focus, and the line DX the directrix of this parabola (by Def. III. and Cor. 2. to Def. VI. of this Book) then because the whole string GPF is equal to the side of the ruler EG; the part of it PF, viz. the distance of the pin from the focus of the parabola, will be equal to the perpendicular PE drawn from the pin to the directrix; therefore the pin P will always move in the parabola OAS (by Cor. 2. XII. of this Book) and therefore by its motion will describe it.

Secondly, let VL be the line given in position bisecting the given line MN in Y, but not at right angles to it, and let LV be the given point in the line VL.

Find QR a third proportional to YV, YM, and from the point V in the line VL take the segment VK equal to a fourth part of QR, and on the side of the point V opposite to where the point Y is: and draw through V, VT parallel to MN, and the right line VF equal to VK, making the angle TVF equal to the angle TVK, and through K draw DX perpendicular to VL: then fix one extremity of the string in the point F, and by means of the pin and ruler let a parabola be described, as in the former case, whose directrix is the line DX and focus F, and this will be the parabola required.

For

For it will pass through the point V, because VF, VK are equal (by Cor. 2. XII. of this Book) and the line VL perpendicular to the directrix will be a diameter (Cor. 1. Def. VI. of this Book) and four times VK its parameter (Cor. 2. XXV. of this Book) and the line VT touches it (by XV. of this Book) because the angles TVF, TVK are equal; and therefore the line MN parallel to this tangent is ordinately applied to the diameter VL, and because the square of YM or YN, by construction, is equal to the rectangle under the abscifs YV and the parameter of the diameter VL, the points M and N are in this parabola (by Cor. 1. Prop. I. of this Book.) *Q. E. F.*

Cor. 1. Hence, if the directrix DX be given in position, and the vertex of the axis given, viz. the point A; the parabola may be described by drawing AD perpendicular to the directrix, and making AF equal to AD; for the point F will be the focus: and likewise if the vertex A and focus F be given; join AF and produce it to D, so that AD be equal to AF, the line drawn through the point D perpendicular to DA will be the directrix. But if the axis AB be given in position, and its vertex A and parameter QR given, let the focus be found by its Definition; likewise the axis being given in position, and the focus and parameter to the axis given; the directrix may be found; therefore in these cases a parabola may be described, as is demonstrated by this Prop.

Cor. 2. Hence likewise, if a diameter EV of a parabola be given in position and its vertex V given, and its parameter be given in magnitude, viz. QR, and a point P in the parabola; this parabola may be described. For in the line VL take the segment VK equal to a fourth part of QR, and through the point K draw the right line KE perpendicular to VL, and through P, PE perpendicular to KE: then let two circles be described from the centers V, P, and with the intervals VK and PE, meeting each other in the points F, G; if a parabola be described, whose directrix is KE and focus F or G, it will pass through the points V and P (Cor. 2. XII. of this Book) and VL will be a diameter (Cor. 1. Def. VI.) and VK taken four times, that is, QR will be the parameter of that diameter FIG. 14.

meter (Cor. 2. XXV. of this Book). It is manifest that in this case two parabolas may be described, which shall solve the problem and one only, if the circles described meet in one point only; but if these circles do not meet each other, the problem in that case will be impossible.

PROP. XXXIV. PROBL. IV.

Two right lines bisecting each other being given in position and magnitude; to describe an ellipse in which the given lines shall be conjugate diameters.

FIG. 25. FIRST, let AB, MN be the given lines bisecting each other in the point C at right angles: and because in this case, AB, MN must be unequal (Cor. 2. XXXI. Book I.) describe a circle from the center M, the extremity of the lesser line MN with an interval equal to CA half of the greater line AB, and cutting AB in the points F, O, and in these points let the end of a string be fixed, of the same length with AB, and by means of the pin P let the string be stretched, and the pin carried round till it returns again to the same point it moved from; the line described by this motion will be an ellipse, in which AB, MN shall be conjugate diameters.

For let an ellipse pass through the points A, M, B, N, of which AB, MN are conjugate diameters (by Prop. XXXIV. Book I.) and because these conjugate diameters are at right angles to each other, the greater of them AB will be the transverse axis and the lesser MN the conjugate axis; and the points F, O the foci, by construction, and Cor. 1. Def. V.

Then because the sum of the lines which are drawn from the pin P to the foci is always equal to AB, the pin P as it is carried round will be in an ellipse (by Cor. 2. XIV. of this Book) and consequently by its motion describes an ellipse.

Secondly, let the two given lines bisecting each other be not at right angles. It is evident that two ellipses cannot have the same center, nor the same two conjugate diameters given in position and magnitude; for if they could, every point in one ellipse would be also

alſo in the other (by Cor. XXXIV. Book I.) that is, the ellipſes would coincide; therefore two conjugate diameters being given in poſition and magnitude the ellipſe will be given, and conſequently the axes will be given in poſition and magnitude, and theſe being found (by Prop. VII. of this Book) the ellipſe may be deſcribed, as in the preceding caſe.

PROP. XXXV. PROBL. V.

Two right lines, which biſect each other, being given in poſition and magnitude; to deſcribe oppoſite hyperbolas of which the given lines ſhall be conjugate diameters.

FIRST, let the two given lines AB, MN biſect each other in the point C at right angles, join AM, and from the point C, in the line AB produced both ways take the ſegments CF, and CO each equal to AM; at the point O let the end of the ruler OG be fixed, ſo that it may be freely carried round this point as a center: and let the end of a ſtring be affixed to the other extremity G of the ruler, the length of which the ruler exceeds by a right line equal to AB, and let the other end of the ſtring be fixed in the point F, and apply the ſtring by means of the pin P to the ſide of the ruler OG, and let the ruler be moved about the point O, and at the ſame time the ſtring, by means of the pin, be conſtantly applied and kept cloſe to the ruler; the pin P by its motion will deſcribe one of the oppoſite hyperbolas, of which the lines AB, MN are conjugate diameters. Fig. 16.

For let the oppoſite hyperbolas be conceived to be deſcribed (by XXXV. Book I.) of which AB may be a tranſverſe diameter and MN its conjugate; the point C will be the center, and becauſe theſe conjugate diameters cut each other at right angles, AB will be the tranſverſe axis and MN the conjugate axis, and conſequently the points F, O will be the foci of the hyperbolas by conſtruction, and Cor. 2. Def. V. but becauſe the exceſs of the ruler GPO above the ſtring GPF is equal to AB, the exceſs of the line PO above PF

will

will be equal to the transverse axis AB; therefore as the ruler is moved the pin P will be in one of the hyperbolas (by Cor. 4. XIV. of this Book). But if the extremity of the same ruler which was affixed to the point O, be now affixed to the point F, and the end of the string be fixed in the point O, and the same as before effected: an hyperbola opposite to the hyperbola now described, will be generated, as is evident.

Secondly, if the two given lines do not cut each other at right angles: two conjugate diameters of an hyperbola being given in position and magnitude, the hyperbolas will be given, and their axes will be given in position and magnitude, as in the second case of the preceding Prop. therefore the axes being found (by Prop. VII. of this Book) the hyperbolas may be described, as in the preceding case.

Cor. *To this and the preceding Proposition.* Hence, if the transverse axis AB of an ellipse or hyperbola be given in position and magnitude, and the foci F, O be given; the ellipse or hyperbola may be described. Or if any diameter AB be given in position and magnitude, and the line PE, which from the given point P in an ellipse or hyperbola is ordinately applied to the diameter AB; an ellipse or hyperbola may be described: for bisect AB in C, and draw through the point C a line parallel to PE, in which line take CM, CN equal to each other, that the square of CA be to the square of CM or CN, as the rectangle AEB to the square of PE: and by the means of the preceding Propositions describe an ellipse or hyperbola, in which AB, MN shall be conjugate diameters; it will pass through the point P, by Cor. XXXIV. and XXXV. Book I.

PROP.

PROP. XXXVI. PROBL. VI.

The directrix and one of the asymptotes of an hyperbola being given in position and the focus given, nearer to this directrix; to describe the hyperbola.

LET DX be the directrix of an hyperbola, and F the focus nearer to that directrix, and let the line CY be parallel to one of the asymptotes; place a ruler on the same plane with the hyperbola, so that the side HE of it be applied to the line DX, and that the other side EG be on that part of DX where the point F is, and let the side of the ruler EG be inclined to the side HE, that it may be parallel to CY: then by means of the pin P, and of the string applied to the side EG, the extremity of which is fixed in the point F, let all be effected as in the description of the parabola, Prop. XXXIII. of this Book; the line described by the motion of the pin P will be an hyperbola, its directrix DX, and asymptote parallel to CY, and focus F. For because the string GPF is equal to GPE, the right line PF drawn from the pin P to the focus is equal to PE drawn from the same pin to the directrix, and parallel to the asymptote; and therefore the pin P moves in an hyperbola, of which DX is the directrix and the point F the nearer focus to this directrix, and whose asymptote is parallel to CY (by Cor. Prop. XIII. of this Book). Fig. 27.

Cor. If a plane cutting a conical surface and making the section of a parabola, be inclined in the least so as to meet the opposite surface, this section will be changed into an hyperbola; and the same analogy may be observed between the descriptions of these sections upon a plane; for if the side of the ruler to which the string is applied in the description of a parabola, be in the least inclined to the other side of the ruler, the pin will describe an hyperbola, and beside when in the description of these sections the side of the ruler passes through the focus, the part PF of the string between the focus and pin will be a fourth part of the parameter of the axis in both sections, as is evident from the Definition of the focus of a parabola, and from (Cor. Prop. XXVI. of this Book).

PROP. XXXVII.

Fig. 28, 29, 30. Let GVH be a right cone, and PAR a conic section in its surface, and LNO a circle which does not meet the section: let its distance AL from the vertex of the section be equal to AF the distance of the same vertex from the focus F nearest to this circle; I say that the intersection of the plane of this circle with the plane of the section will be its directrix, and that PN a side of the cone intercepted between this circle and any point P in the section will be equal to a right line drawn from the same point to the focus F nearest to this circle.

LET the cone be cut through its axis VX by a plane GVH perpendicular to the plane PAR of the section, and cutting it in the right line AFB, this line will be the axis of the section, which in the ellipse or hyperbola will be the transverse, and consequently the vertex A, the focus F of the section, and the line AL will be in the plane GVH: then because the plane of the section PAR, and the plane of the circle LNO are both perpendicular to the plane GVH, their common intersection DE will be perpendicular to the plane GVH (19. 11.) and consequently to the axis AB of the section; and because DE is in the plane of the circle LNO, it will be parallel to the plane of the cone's base; let the line DL be the intersection of the plane GVH with the plane of the circle LNO, it will be parallel to the plane of the base, as is manifest.

Fig. 28. *Case* 1. First, let the section PAR be a parabola, and draw from a point P in it PQ an ordinate applied to the axis, it will be parallel to DE, and therefore to the plane of the base: let a plane pass through PQ parallel to the base, and intersecting the plane GVH in the line GQH, which will be parallel to DL (16. 11.) then because the triangles GAQ, DAL are similar to each other and to the triangle GVH, and the lines GV, HV are equal, because the cone

cone is right, GA, QA, as likewise DA, LA will be equal, and consequently QD will be equal to GL; therefore since DA the distance of the line DE from the vertex of the parabola is equal to AL, that is to AF the distance of the focus from the same vertex, the line DE perpendicular to the axis AB will be the directrix of the parabola; therefore the first part of this Prop. is evident when the section is a parabola. Wherefore if a right line PF be drawn from the point P to the focus, it will be equal to the perpendicular drawn from the same point P to the directrix (by XII. of this Book) that is, to QD or GL; but GL, PN are equal, being segments of the sides of a right cone intercepted between parallel circles, and therefore PN is equal to PF.

Case 2. Let the section PAR be an ellipse and C the center, and draw the conjugate semiaxis CP, it will be parallel to the line DE, and therefore to the plane of the base; let a plane pass through the line CP, parallel to the base, and intersecting the plane GVH in the line MCK, which is parallel to DL (16. 11.) MK will be a diameter of the circle MPK, and by a property of the circle, the rectangle MCK is equal to the square of the conjugate semiaxis CP; draw from the vertex V of the cone a line parallel to AB, and meeting the diameter GH of the base, produced in S: then by Prop. X. Book I. the rectangle ACB, or the square of CA, is to the rectangle MCK, or the square of CP, as the square of VS to the rectangle HSG; therefore, by conversion, the square of CA is to its excess above the square of CP, that is, to the square of CF (Cor. 1. Def. V. of this Book) as the square of VS, to its excess above the rectangle HSG, that is, to the square of VG*: but because the triangles VSG, ACM are similar, the square of CA is to the square of AM as the square of VS to the square of VG; there- FIG. 29.

* *The square of VG.*] The triangles VXS, VXG (Fig. 29, 30.) are right angled, and therefore the difference between the squares of the hypotenuses VS and VG is equal to the difference between the squares of the sides XS and XG, that is, to the rectangle HSG under GS, the sum of these sides, and HS their difference. So that in Fig. 29, the square of VG is the excess of the squares of VS above the rectangle HSG: and in Fig. 30, the square of VG is equal to the square of VS and the rectangle HSG.

 fore

fore the lines CF and AM are equal; but because MC, DL are parallel, MA or CF, the distance of the focus from the center, will be to the semiaxis CA, as LA or FA to DA, the distance of the line DE from the vertex A of the ellipse; therefore the line DE, being perpendicular to the axis, will be the directrix (by Cor. 2. XI. of this Book) and consequently the first part of the Prop. is evident when the section is an ellipse. Now draw from any point P in the ellipse the right line PC perpendicular to the axis AB, and let the plane of a circle pass through PC, and meet the sides VG, VH in the points M, K, and intersect the plane GVH in the line MK parallel to DL; then because CD is equal to the perpendicular drawn from P to the directrix DE, if PF be drawn to the focus (by XI. of this Book) PF will be to CD, as FA or LA to DA: but ML is likewise to CD as LA to DA; therefore PF, ML are equal; but ML, PN are equal, as in the case preceding; therefore PN is equal to PF.

FIG. 30. *Case* 3. Let the section PAR be an hyperbola, and let PR be the intersection of its plane with the plane of the base HPG, it will be parallel to DE (16. 11.) and therefore ordinately applied to the transverse axis AB, which produced let it meet in Q: let HQG be the intersection of the plane HVG with the plane of the base, it will be parallel to DL (16. 11.) draw from the center C of the hyperbola a line parallel to HQG, DL, meeting the sides HV, GV of the cone in the points K, M, and draw from the vertex of the cone VS parallel to the axis AB of the hyperbola, meeting HQG in S, and let CT be the conjugate semiaxis.

Because by Prop. X. Book I. the rectangle ACB, that is, the square of CA is to the rectangle KCM, as the square of VS to the rectangle HSG, and in like manner the rectangle AQB (under the abscisses) is in the same ratio to the rectangle HQG, or to the square of the ordinate PQ, the rectangle KCM will be equal to the square of the conjugate semiaxis CT, (by Def. XXIV. Book I.) and therefore, by conversion, the square of CA is to the sum of the squares of CA and CT, that is, to the square of CF (by Cor. 2. Def.

Def. V. of this Book) as the ſquare of VS to the ſum of the ſquare of VS and the rectangle HSG, that is, to the ſquare of VG *: but becauſe the triangles CAM, SVG are ſimilar, the ſquare of CA is to the ſquare of MA, as the ſquare of VS to the ſquare of VG; therefore the lines CF, MA are equal: but becauſe MC, DL are parallel, CA is to MA or CF, as DA to LA or AF, therefore the line DE is the directrix of the ſection (Cor. 2. XI. of this Book) wherefore the firſt part of this caſe is manifeſt.

Now draw from any point in the hyperbola a right line PF to the focus, and PQ perpendicular to the axis AB of the hyperbola, and draw through PQ the plane of the circle GPHR, it may be ſhewn in the ſame words as in the preceding caſe, that the line PF is equal to GL and conſequently to PN. *Q. E. D.*

Cor. As there are two foci F, and *f* in an ellipſe and hyperbola, there will be two circles LNO, *lno*, ſuch as are deſcribed in the propoſition, whoſe planes will interſect the plane of the ſection in the lines DE, *de*, which will be directrices of the ſection; and becauſe the lines PF, P*f* drawn to the foci are equal to PN, P*n*, the ſegment N*n* of any ſide of the cone intercepted between the circles LNO, *lno*, will be equal to the tranſverſe axis of the ſection, by Prop. XIV. of this Book.

* See the note in page 123.

END OF THE SECOND BOOK.

TAB. 9.
Page. 126.

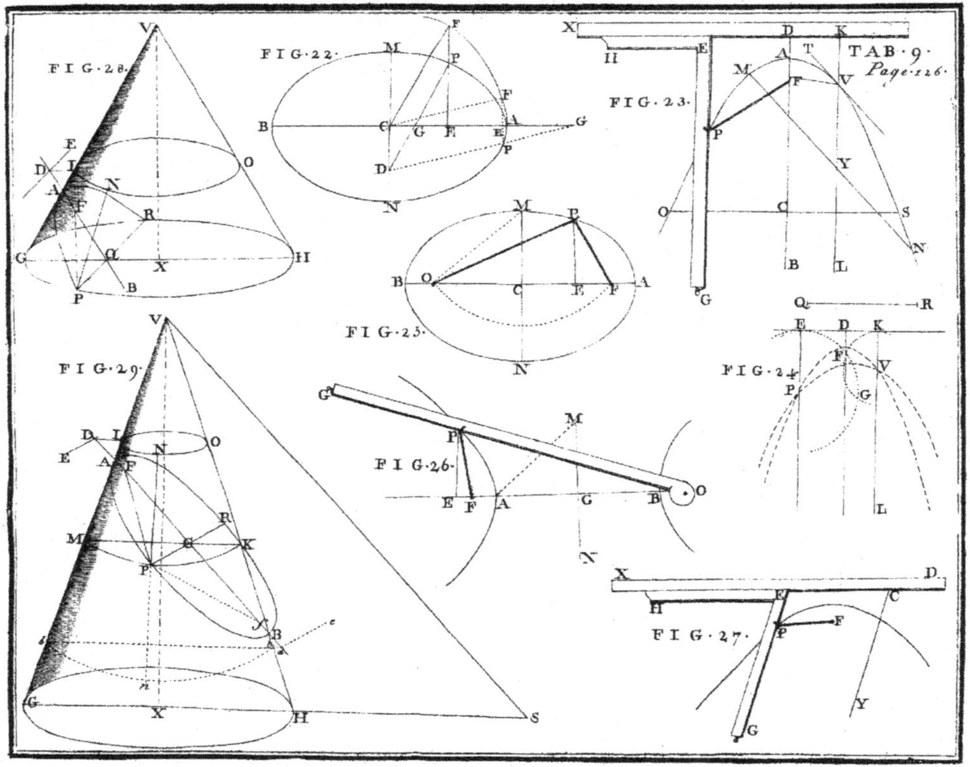

CONIC SECTIONS.

BOOK THE THIRD.

Of the Parabola.

PROPOSITION I.

If a right line FG touching a parabola in F meets two diameters VC, DH in the points G and R; the squares of the segments FG, FR of the tangent between the point of contact F and the diameters, will be to each other as the segments GV, RD of the diameters between their vertices and the tangent. Or if a right line AB cutting a parabola, meets two diameters FL, VC in the points K, E; the rectangles AKB, AEB will be to each other as the segments FK, VE of the diameters between their vertices and the secant AB. Fig. 1.

Part

Part 1. LET the right line P be the parameter of the diameter which passes through the point of contact F, the square of FG is equal to the rectangle under GV and the line P, and the square of FR is equal to the rectangle under RD and the line P (by Prop. II. Book II.) therefore the squares of FG, FR are to each other as the segments GV, RD.

Part 2. Let the right line P be the parameter of the diameter to which the secant AB is ordinately applied; the rectangle AKB is equal to the rectangle under FK and the line P; and the rectangle AEB is equal to the rectangle under VE, and the line P (Prop. III. Book II.) and therefore the rectangle AKB, AEB are to each other as the segments FK, VE. *Q. E. D.*

Cor. Let a right line meet a parabola in the points A, B, and let F be the vertex of the diameter meeting the right line AB in the point K; if from any other point E in the line AB, EV be drawn parallel to FK, so that it be to FK as the rectangle AEB to the rectangle AKB; the point V will be in the parabola which passes through the points A, F, B, and whose diameter is FK. When the point E is between A and B, the line EV must be drawn on the same side of AB with the point F, and when not, on the contrary side.

PROP. II.

If from the vertices of two diameters ordinates be drawn applied to these diameters; the abscisses between the ordinates and vertices of the diameters will be equal.

Fig. 21. FROM the vertices of the diameters FG, HM draw the ordinates FL, HK applied to the diameters HM, FG; the abscisses HL, FK will be equal.

Draw through the vertex of the diameter HM a right line touching the parabola and meeting the diameter FG in A; then because HAFL is a parallelogram, the abscis HL is equal to AF, that is, to the abscis FK, by Prop. XLVII. Book I. *Q. E. D.*

PROP.

PROP. III.

If one side of any triangle be parallel to the diameters of a parabola; the squares of the other sides will be to each other as the parameters of the diameters, whose ordinates are parallel to those sides.

LET the side MN of the triangle LMN be parallel to the diameters of the parabola BFH, and let the right lines P and Q be the parameters of the diameters ED, FR, to which BC, BR (parallel to LM, LN) are ordinately applied; the square of LM will be to the square of LN as P to Q. FIG. 3.

For through the vertices E and F of the diameters ED, FR, draw the right lines EG, FG touching the parabola; they will be parallel (by hypothesis, and Cor. 8. XXV. Book I.) to LM, LN; let the tangent EG meet the diameter FR in A, and from the point of contact E, draw EK an ordinate applied to the diameter FR, it will be parallel to FG, and (by XLVII. Book I.) KF, FA will be equal; therefore EG, GA are equal: then because the sides of the triangles LMN, GAF are respectively parallel, those triangles will be equiangular; therefore the square of LM is to the square of LN, as the square of (GA or) GE to the square of GF, that is, as the parameter P to the parameter Q, by Cor. 1. Prop. III. Book II. *Q. E. D.*

COR. 1. If from any point B in a parabola an ordinate BC be drawn to a diameter ED, and any other right line BD to the same diameter; the square of BD will be equal to the rectangle under the abscis EC, and the parameter of the diameter FR, to which the line, parallel to BD, is ordinately applied. For let Q be the parameter of the diameter FR, and P the parameter of the diameter ED: by this Prop. the square of BC is to the square of BD as (P to Q or as) the rectangle under P, EC to the rectangle under Q, EC: but (by Prop. I. Book II.) the square of BC is equal to the rectangle under P, EC; therefore the square of BD is equal to the rectangle under Q, EC.

Cor. 2. If a right line OE touches a parabola, and through a point O taken in it a right line be drawn cutting the parabola in B and H, and meeting in D the diameter drawn through the point of contact E; the square of the segment OD between the tangent and the diameter will be equal to the rectangle BOH under the segments of the secant between the tangent and parabola. For, by this Prop. the square of OD is to the square of OE, as Q to P, and (by Cor. 3. III. Book II.) the rectangle BOH is in the same ratio to the square of OE, therefore the square of OD is equal to the rectangle BOH.

Cor. 3. The parameter of any diameter of a parabola is to the parameter of its axis, as the square of radius to the square of the sine of the angle which that diameter contains with its ordinates, as is evident from this Prop. and because the squares of the sides of a triangle are as the squares of the sines of the opposite angles; and therefore the parameters of the diameters of any parabola are reciprocally, as the squares of the sines of the angles which those diameters contain with their ordinates; and therefore the principal parameter, that is, the parameter of the axis is less than any other parameter.

PROP. IV.

If the right line joining the extremities of two ordinates meets the diameter to which they are applied: the part of the diameter intercepted between its vertex and the right line joining the extremities of the ordinates, will be a mean proportional between the abscisses.

Fig. 4. LET EB, GD be ordinates applied to the diameter AD of the parabola, and let a line joining the points E, G meet the diameter AD in C; CA will be a mean proportional between the abscisses AB, AD.

Draw through the points E, G two tangents which may meet the diameter AD in H and F; AH, AB, as likewise AF, AD will be equal (XLVII. Book I.) but it is evident (from Prop. LIV. Book I.) that

that AC between the vertex of the diameter and the line joining the points of contact, is a mean proportional between AH and AF; and therefore is a mean proportional between AB and AD. *Q. E. D.*

COR. 1. Hence, if two right lines be drawn from a point E in a parabola, one of them a tangent, the other cutting the section in the points E, G, which shall meet any diameter AL in H and C; the segments EC, CG of the secant between the diameter and parabola will be to each other as the segments AH, AC of the diameter between the vertex and tangent, and the same vertex and secant. For if EG be an ordinate applied to the diameter AL, EG will be bisected in C and HC in A; but if EG be not an ordinate to AL, draw from the points E, G the ordinates EB, GD to the diameter AL: and because the triangles EBC, DGC are similar, the square of EC is to the square of CG as the square of EB to the square of GD (that is, as the abscis AB to the abscis AD, or because AB, AC, AD are proportional) as the square of AB to the square of AC; therefore EC will be to CG, as (AB or) AH to AC.

PROP. V.

Let the right line HK be an ordinate to the diameter FP: through its extremity H draw the diameter HC, and from the vertex of the diameter FP draw a right line meeting the parabola again in O, and the diameter HC in L, and HK in N: FO, FL, FN will be proportional. FIG. 5.

FOR draw from the points O, L, two lines OP and LS, to the diameter FP, and parallel to the ordinate HK; OP will be an ordinate applied to the diameter FP, and LS will be equal to HK: then because OP, LS are parallel, the square of FO will be to the square of FL as the square of OP to (the square of LS or) HK, that is, as the abscis FP to the abscis FK, or because OP and HK are parallel, as FO to FN; therefore FO, FL, FN are proportional. *Q. E. D.*

SCHOLIUM.

FIG. 5. *If two indefinite right lines DE, CM meet each other in a given angle in the point H, and about any point F another indefinite right line revolves meeting DE, CM in the points L, N, and a point O be taken in the revolving line on the same side of the point F where N is (the intersection of the revolving line with the line DE) so that FO, FL, FN be continually proportional; the* Locus *of all the points O will be a parabola passing through the points F and H, and to which CM will be a diameter, and DE will be parallel to the ordinates applied to the diameter which passes through the point F: but if a mean proportional FO be taken in this revolving line between FL, FN; the line described by the point O will be an hyperbola. See the Scholium to* Prop. XLVI. Book I.

PROP. VI.

If three right lines touching a parabola meet each other; they will be cut in the same ratio, viz. between their points of concourse and points of contact.

FIG. 6, 7. LET ABC be a parabola, which the lines AD, FE, DC touch in the points A, B, C; AF will be to FD as FB to BE, and as DE to EC. For let the point of contact B in the segment of the parabola fall between the points A, C: join AC and bisect it in G, the diameter drawn through G will pass through the point of concourse of the tangents AD, CD (Cor. 1. XXVI. Book I.)

FIG. 6. *Case* 1. If the diameter DG passes through the point B, the tangent FBE will be parallel to AC, and therefore will be bisected in the point B; and because GB, BD are equal, and AC, FE parallel, the tangents AD, CD will be bisected in the points F, F.

FIG. 7. *Case* 2. But if the diameter DG does not pass through the point B; join AB, BC, and draw through the points F, B, E the diameters FHL, BN and EKM; and (by Cor. 1. XXVI. Book I.) AH, HB will be equal, and consequently AL, LN; and by the same

ſame reaſoning NM, MC will be equal; therefore LM is equal to (half of AC, that is) to GC, and therefore (taking away the common part GM) LG will be equal to MC or NM, and therefore LN will be equal to GM: theſe being premiſed, AF will be to FD as AL to LG (2. 6.) that is, as LN to NM, or GM to MC; but as LN is to NM, ſo is FB to BE, and as GM is to MC, ſo is DE to EC; therefore as AF is is to FD, ſo is FB to BE, and DE to EC. *Q. E. D.*

PROP. VII. PROBL. I.

The directrix DX of a parabola being given in poſition, and the focus F given, and the right line MN, not parallel to the diameters, given in poſition; to draw a tangent to the parabola, which ſhall be parallel to this right line. FIG. 8.

FROM the focus F draw FO perpendicular to MN, meeting the directrix in E, and FE being biſected in T, draw TP parallel to MN, which may meet in P the perpendicular to the directrix drawn through E; and PF being joined, the triangles PTE, PTF will be equal (4. 1.) and therefore ſince FP is equal to PE, the point P will be in the parabola (Cor. 2. XII. Book II.) and becauſe the angles EPT, FPT are equal, the line PT parallel to MN touches the parabola. *Q. E. F.*

PROP. VIII. PROBL. II.

Through three given points to deſcribe a parabola, the diameters of which ſhall be parallel to a right line given in poſition, provided it be not parallel to a right line joining any two of the given points.

LET B, G, F be the given points but not in the ſame line, and let QR be the line given in poſition: join GF, and draw through the point B a line parallel to QR, meeting GF in E: if the line GF be biſected in E, the propoſition is the ſame with FIG. 9.

Prop.

Prop. XXXIII. Book II. But if GF be not bisected in E, let it be bisected in M, and let the line BE be to VM, drawn through M parallel to BE, as the rectangle GEF to the rectangle GMF; then by Prop. XXXIII. Book II. describe a parabola, of which the line VM shall be a diameter and V its vertex, and to which GF shall be ordinately applied: this parabola will pass through the point B (by Cor. Prop. I. of this Book) and its diameters are parallel to QR by construction. When the point E falls between G and F, the line VM must be taken on the same side of GF, where the point B is; but if the point E be not between G, F, then on the contrary side. Cor. Prop. I. of this Book. *Q. E. F.*

PROP. IX. PROBL. III.

Four points being given in a parabola; to describe the parabola.

FIG. 9. *Case* 1. LET A, B, G, F be the given points, and let the right lines joining these points form a trapezium, having none of its sides parallel: let two of the opposite sides produced meet in D, and as the rectangle BDA is to the rectangle GDF, so let the square of DB be to the square of the segment DE of the line DGF; BE being joined, describe a parabola (by Prop. preceding) passing through the points G, B, F, whose diameters are parallel to BE; it will pass through the point A.

For because the rectangle BDA and the square of DB are unequal, the rectangle GDF and the square of DE will be unequal, by construction; therefore the line DB does not touch the parabola (as is evident from Cor. 2. III. of this Book) and therefore must cut it in another point, and if not in A, suppose in X; then (by Cor. 2. Prop. III. Book II.) the rectangle BDX is to the rectangle GDF as the square of DB to the square of DE (by III. of this Book) that is, by construction, as the rectangle BDA to the rectangle GDF; therefore the rectangles BDX, BDA are equal, and therefore the lines DA, DX are equal, which is absurd (for the point X is on the same side of D as the points B and A, because D is without

out the parabola) therefore the parabola meets the line DB in the point A. Because the segment DE in the line DGF may be taken on either side of the point D, two parabolas may be described, which will answer the problem.

Case 2. Let B, G, F, P be the four given points, and let the right lines joining the points G, F and B, P be parallel: the right line MN bisecting these parallels will be a diameter of the parabola passing through these four points (by Cor. 6. XXV. Book I.) let then a parabola be described passing through the three points B, G, F, whose diameters shall be parallel to the line MN, and since the line BN is ordinately applied to a diameter of this parabola and PN is equal to BN, this parabola will pass through the point P; it is manifest in this case that there can be but one position of the diameters of the parabola which passes through these points, and consequently one parabola only can pass through these points.

Cor. From the construction a method of finding the position of the diameters of a parabola from four points given in it, is manifest.

PROP. X.

Two parabolas, having a common axis and the same principal parameter, produced indefinitely, continually approach each other, but never meet.

LET AEF, DCB be two parabolas having the common axis EG, and let the right line PQ be the principal parameter of both: let an ordinate be applied to the axis and meet it in G, the exterior parabola in the points A, F, and the interior in D. Because the square of AG is equal to the rectangle under GE, PQ, the difference of these squares, that is, the rectangle ADF (5. 2.) will be equal to the rectangle under CE, PQ; then because the rectangle CE, PQ is given, the rectangle ADF will be given in magnitude: but the farther the point D recedes from the vertex C of the interior Fig. 101.

terior

terior parabola, the greater the line DF will be, and therefore the diftance AD between the parabolas will be lefs: but the points A, D can never coincide, becaufe the rectangle ADF is always a given magnitude. *Q. E. D.*

Cor. Hence, the fegment DH intercepted between the vertices of two coinciding diameters is equal to the fegment EC between the vertices of the axes. For (by III. Book II.) the rectangle ADF is equal to the rectangle under the lines PQ, DH, and by this Prop. the fame rectangle ADF is equal to the rectangle under PQ, EC; therefore the fegments DH, EC are equal.

The QUADRATURE of the PARABOLA.

PROP. XI.

Fig. 11. If a right line BC be infcribed in a parabola, and tangents be drawn through the points B, C, meeting each other in the point A, and if the line BC be bifected in G and its parts be again bifected, and in like manner the bifections of the parts be ftill continued; and if diameters be drawn through all the bifecting points D, E, V, G, &c. meeting the parabola in L, M, N, O, &c. and if all thefe points B, L, M, N, O, &c. be joined by lines which may form an infcribed figure, and if tangents be drawn through the fame points B, L, M, N, O, &c. which may form a circumfcribed figure: the figure infcribed will be always double the area contained within the triangle BAC and without the circumfcribed figure.

FOR

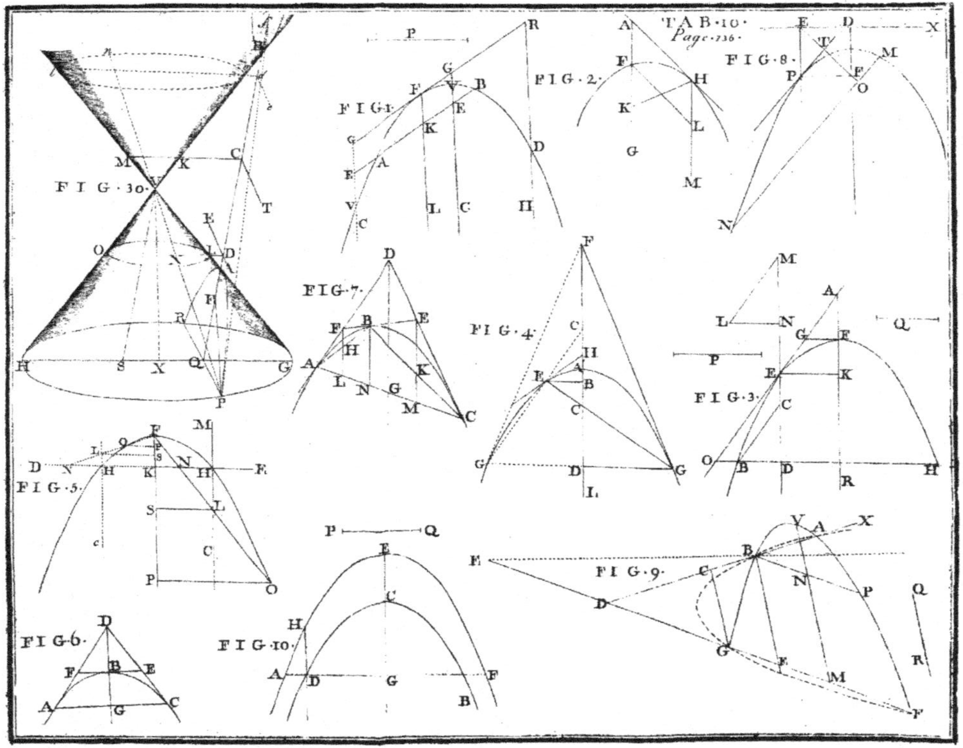
TAB·10·
Page·236·
FIG·1·
FIG·2·
FIG·3·
FIG·4·
FIG·5·
FIG·6·
FIG·7·
FIG·8·
FIG·9·
FIG·10·
FIG·30·

FOR first let the inscribed figure be the triangle BOC, the circumscribed figure will be the trapezium BHZC, and the area without it, will be the triangle HAZ: and because GO is a diameter, it will pass through the point A (Cor. 1. XXVI. Book I.) and because GA is double OA (XLVII. Book I.) BC will be double of HZ; therefore the triangle BOC is double the triangle HAZ; then if BG be bisected in E, the diameter EM will bisect BO, and consequently will pass through the point H; therefore, because FM, MH are equal, the triangle BMO will be double the triangle KHI: but BMO is added to the first inscribed figure, and KHI is added to the area, which is without the circumscribed figure: and in like manner, if the number of sides be increased in the inscribed and circumscribed figures, whatsoever is added to the area without the circumscribed figure, the double of it is always added to the inscribed figure; and therefore the inscribed figure will be always double the area which is without the circumscribed figure. *Q. E. D.*

Cor. 1. Hence, it is evident, that no figure can be inscribed (as in the proposition) in the parabolic segment BOQC which can be double the area BACO, without the parabolic segment and within the triangle BAC.

Cor. 2. It is also evident that no figure can be circumscribed about the parabolic segment BOQC (as in the proposition) so that the area without it and within the triangle shall be half the parabolic segment BOQC.

PROP. XII.

The above construction remaining, a figure may be inscribed (as in the preceding Proposition) in the parabolic segment BOQC, terminated by the parabola and a right line, which shall not be less than the segment by any given minute space whatsoever: Or a figure may be circumscribed about this parabolic segment (as in the preceding Proposition) Fig. 11.

S and

and within the triangle BAC, which ſhall cut off from the area BACO a ſpace leſs than any given ſpace whatſoever.

Part 1. LET the triangle BOC be inſcribed in the parabolic ſegment, as in the preceding Prop. Since it is half of the whole triangle BAC, it will be greater than the half of the parabolic ſegment BOQC: and in like manner if the triangles BMO and CQO be inſcribed in the remaining ſegments, they will be greater than the halves of theſe ſegments: and in the ſame manner it will be in the remaining ſegments; wherefore a figure may be inſcribed which ſhall not be leſs than the parabolic ſegment by any given ſpace whatſoever (1. 10.)

Part 2. In the area BACO inſcribe the triangle HAZ, whoſe baſe HZ ſhall be parallel to the line BC, joining the points of contact; becauſe GO and OA are equal, BH and HA, and likewiſe CZ and ZA will be equal: therefore the triangle HAZ is half of the triangles BOA and COA together, and therefore is greater than the half of the area BACO; and in like manner, if in the remaining ſegments, BHOM and CZOQ, the triangles KHI, RZP be inſcribed, they will be greater than the halves of theſe ſegments, and if theſe inſcriptions be continued, the ſame will always follow; wherefore the circumſcribed figure at length ſhall cut off from the area BACO a ſpace leſs than any given ſpace whatſoever.

PROP. XIII.

FIG. 11. **Let BOQC be a ſpace terminated by a parabola and the right line BC: draw a line parallel to BC, touching the parabola in O, and meeting in X and T, the diameters drawn through the points B and C; the parabolic ſegment BOQC will be to the circumſcribed parallelogram BXTC as 2 to 3.**

DRAW the tangents BA, CA, and through their point of concourſe A the diameter AG, which will paſs through the point of contact O, and will biſect BC in G; then becauſe AG is double of of AO, the triangle BAC will be equal to the parallelogram BXTC; but the parabolic ſegment BOQC is double the area BACO; for if it exceeded the double of that area, by any ſpace S, then by the firſt part of the preceding Prop. a figure may be inſcribed in the ſegment BOQC, which ſhall differ leſs from this ſegment by a ſpace leſs than S, and therefore this figure will be greater than the double of the area BACO, contrary to Cor. 1. Prop. XI. and therefore the ſegment BOQC does not exceed the double of the area BACO; nor is it leſs than the double of this area; for if the area BACO could exceed the half of the ſegment BOQC, by any ſpace whatſoever as S; then by the ſecond part of the preceding Prop. a figure may be circumſcribed about this ſegment which ſhall cut off a ſpace leſs than the ſpace S; and therefore the area without this figure and within the triangle BAC would exceed the half of the ſegment BOQC, contrary to Cor. 2. Prop. XI. therefore ſince the ſegment BOQC does not exceed the double of the area BACO, nor is leſs than the double, it neceſſarily muſt be double of this area, and therefore is to the whole triangle BAC, that is, to the parallelogram BXTC as 2 to 3. *Q. E. D.*

Cor. 1. Two ſegments LDP, AKE of the ſame parabola are to one another in the *ſeſquiplicate* ratio of the abſciſſes DM, KT of the diameters, which biſect the lines LP, AE. Draw from the points L and A in the parabola the lines LR, AH perpendicular to the diameters DM, KT, they will be the altitudes of the parallelograms LNDM, ABKT; therefore LNDM is to ABKT in a ratio compounded of the ratio of DM to KT and of LR to HA; but the ſquare of LR is to the ſquare of HA, as DM to KT (by Cor. 1. III. of this Book) wherefore LR is to HA in the *ſubduplicate* ratio of DM to KT, and conſequently LNDM and ABKT are in the *ſeſquiplicate* ratio of DM and KT, and therefore the whole parallelograms circumſcribed about theſe ſegments are in the ſame ratio; but the parabolic ſegments LDP and AKE are to one another as thoſe parallelograms, by the Propoſition, whence the Corollary is evident. Fig. 12.

greater than double that area by any given quantity, ſuppoſe the ſpace S, then ſince a ſeries of triangles might be inſcribed in the ſegment BMOQC, which would differ from it by a quantity leſs than the ſpace S, this ſeries of triangles would be more than double the area BACO, and conſequently would be more than double the ſeries of correſponding triangles inſcribed in the area BACO, contrary to what was proved before. Or if the ſegment BMOQC was leſs than double the area BACO, then that area would be greater than one half of the ſegment BMOQC by a certain quantity, ſuppoſe the ſpace S, and therefore ſince a ſeries of triangles might be inſcribed in the area BACO, which would differ from it by a quantity leſs than the ſpace S, this ſeries would be greater than one half of the parabolic ſegment BMOQC, and conſequently greater than half of the whole ſeries of correſponding triangles inſcribed in that ſegment, contrary to what was proved before. Therefore it appears that the parabolic ſegment BMOQC is neither more nor leſs than double the area BACO, and conſequently the ſegment BMOQC is to the whole triangle BAC in the proportion of 2 to 3. But the triangle BAC is equal to the circumſcribing parallelogram, as having double its altitude and being on the ſame baſe with it, and therefore the parabolic ſegment BMOQC is to the circumſcribing parallelogram BXTC in the proportion of 2 to 3. *Q. E. D.*

END OF THE THIRD BOOK.

CONIC SECTIONS.

BOOK THE FOURTH.

The Ellipse and Hyperbola.

PROPOSITION I.

If through the vertices of two conjugate diameters, four right lines be drawn touching an ellipse or conjugate hyperbolas; the parallelogram formed by the tangents will be equal to the rectangle under the axes.

LET C be the center of an ellipse or hyperbola, AB, Mm the axes, and ED, FG any conjugate diameters; the tangents drawn through their vertices will form a parallelogram (by Cor. XXVII. Book I.) let the tangents drawn through the points F and D, meet each other in H; FCDH will be a fourth part of this parallelogram; draw from the center CK perpendicular to the tangent FH. Fig. 1. 2.

FH. Becauſe (by Cor. 1. XXXI. Book II.) the perpendicular CK is to the tranſverſe ſemiaxis CA, as the conjugate ſemiaxis CM to the ſemidiameter CD parallel to the tangent FH, the rectangle under CK and CD, that is, the parallelogram FCDH will be equal to the rectangle under the ſemiaxis CA and CM; whence the Propoſition is evident.

Cor. It is manifeſt, that a parallelogram inſcribed in an ellipſe or conjugate hyperbolas, by joining the vertices of two conjugate diameters is half of the parallelogram circumſcribed about theſe conjugates, and therefore all ſuch parallelograms are equal.

PROP. II.

Fig. 3. If from the vertices of two conjugate diameters CF, CG of an ellipſe, the ordinates FE, GH be drawn to a third diameter AB; the ſquare of the part, intercepted between either of the ordinates and center, will be equal to the rectangle under the ſegments between the other ordinate and the vertices of the diameter, to which the ordinates are applied.

FOR through the vertex of either of the diameters CF, let a line be drawn touching the ellipſe, and meeting the diameter AB in D, and its conjugate M*m* in O: becauſe FD, CG, and FE, GH are parallel, the triangles FED, GHC will be equiangular, and becauſe CO, FE are parallel, the rectangles OFD, CED will be ſimilar; and therefore OFD is to CED, as the ſquare of FD to the ſquare of ED; or becauſe the triangles are ſimilar, as the ſquare of CG to the ſquare of CH: but the rectangle OFD is equal to the ſquare of CG (LI. Book I.) therefore the ſquare of CH is equal to the rectangle CED, that is, to the rectangle AEB (Cor. 1. XLIX. Book I.) under the ſegments between the other ordinate FE and the vertices of the diameter AB; and theſe equals being taken from

from the ſquare of CB, the remainders, viz. (5. 2.) the rectangle AHB and the ſquare of CE will be equal. *Q. E. D.*

Cor. It is evident that the ſquares of the ſegments of the diameter AB, to which the ordinates are applied, viz. between the ordinates and center, are together equal to the ſquare of the ſemidiameter CB: for ſince the ſquare of CH is equal to the rectangle AEB, the ſquare of CH, together with the ſquare of CE, will be equal to the ſquare of CB (5. 2).

PROP. III.

The ſum of the ſquares of any two conjugate diameters of an ellipſe is equal to the ſum of the ſquares of the axes.

LET CB, CM be the ſemiaxes of an ellipſe, and CF, CG two conjugate ſemidiameters, and let EF, GH be perpendicular to CB, and FL and GN perpendicular to CM. Fig. 3.

Becauſe, by the preceding Cor. the ſquare of CB is equal to the ſquares of CE, CH together, and the ſquare of CM equal to the ſquares of CL, CN, that is, to the ſquares of FE, GH; the ſquares of CB, CM together will be equal to the four ſquares of CE, CH, FE, GH, to which likewiſe the ſquares of CF, CG are equal (47. 1.) therefore the ſquares of CF, CG are equal to the ſquares of CB, CM together, and therefore the ſquares of the diameters are equal to the ſquares of the axes. *Q. E. D.*

PROP. IV.

If the angles contained by the asymptotes of an hyperbola are right angles; any two conjugate diameters will be equal to each other: but if the angles contained by the asymptotes are not right angles, any two conjugate diameters will be unequal, and the difference of their squares will be equal to the difference of the squares of any other two conjugate diameters.

FIG. 4. *Part* 1. LET AD, AE be the asymptotes of an hyperbola, and its center A; draw from the center A any semidiameter AB, and through the vertex B a tangent meeting the asymptotes in the points D, K; BD or BK will be equal to the semidiameter conjugate to AB (XXXVIII. Book I.) therefore if DAK be a right angle, a circle described from the center B about the diameter DK will pass through the point A (Convers. 31. 3.) and therefore the semidiameter AB is equal to BD or BK, that is, to the semidiameter conjugate to it.

In this case, the hyperbola is said to be *equilateral*.

Part 2. Let the asymptotes AD, AE of the hyperbola contain an acute angle, and draw the semidiameter AB and tangent DBK, as before: because DAE is an acute angle, it is without the semicircle described upon the diameter DBK (Convers. 31. 3.) and therefore the transverse semidiameter AB will be greater than BD or BK. Draw any other semidiameter AC, and a tangent through its vertex C meeting the asymptotes in the points F, E: it is required to shew that the difference of the squares of AB, BD is equal to the difference of the squares of AC, CF. From the points B, K, C, E, draw BG, KL, CH and EM perpendicular to the asymptote DA: because DB, BK are equal, DG, GL will be equal, and because FC, CE are equal, FH, HM will be equal, and since BG is a perpendicular, therefore the difference of the squares of AB, BD will be equal to the difference of the squares of GA, GD or GL, that is, is, to the rectangle DAL (6. 2.) and in like manner the difference of the squares of AC, CF is equal to the difference of the squares of HA, HF or HM, that is, to the rectangle FAM: but because the triangles DAK, FAE are equal (Cor. 2. XLV. Book I.) the sides about the common angle are reciprocal, that is, DA will be to FA (as AE to AK, that is, because of the parallels) as AM to AL; therefore the rectangle DAL is equal to the rectangle FAM; and therefore the difference of the squares of AB, BD is equal to the difference of the squares of AC, CF. It is evident that in an hyperbola conjugate to BCN, every transverse diameter is less than its conjugate diameter. *Q. E. D.*

If the angle contained by the asymptotes be half a right angle, the excess of the squares of two conjugate diameters will be double the parallelogram described about them; and if the axes of an hyperbola and their difference be proportionals, that excess will be equal to that parallelogram.

PROP. V.

The two diameters of an ellipse, which bisect the right lines joining the vertices of the axes, are conjugate diameters and equal to each other: and on the contrary, if two conjugate diameters of an ellipse be equal, they will bisect the right lines joining the vertices of the axes.

Part 1. LET AB, M*m* be the axes of an ellipse, and join AM, BM, and draw the diameters ED, FG, which bisect these lines, and let FG meet BM in I, and because MI, IB, and also AC, CB are equal, the diameter GF will be parallel to AM, which is bisected by ED; therefore ED, FG are conjugate diameters (XXVII. Book I.) and because MCB is a right angle, and IB, IM equal, the center of the circle described about the triangle MCB will be in the point I; therefore IC will be equal to IB, and conse- Fig. 5.

consequently the angle ICB is equal to the angle IBC, that is, to the alternate angle BCD; therefore the diameters ED, FG are equal, by Cor. 3. Prop. VIII. Book II.

Part 2. If two conjugate diameters ED, FG of an ellipse be equal, and from the extremity B of one axis a line be drawn parallel to either of the conjugate diameters, suppose to ED, it will pass through the extremity of the other axis; for let this line meet the ellipse again in M, and the diameter FG in I, BM will be bisected in I (Cor. XXVII. Book I.) join CM, and because the diameters ED, FG are equal, the angle ICB is equal to the angle BCD (Cor. 3. VIII. Book II.) that is, to the alternate angle IBC, and consequently IC is equal to IB or IM; therefore a circle may be described from the center I passing through the points B, C, M; and therefore the angle MCB in a semicircle will be a right angle; wherefore since AB is an axis, MC*m* will be the other axis; and therefore the equal conjugate diameters ED, FG bisect the lines joining the vertices of the axes. *Q. E. D.*

Cor. 1. Hence it is manifest that there can be but two conjugate diameters of an ellipse equal to each other.

Cor. 2. Hence likewise, if the lines MA, MB be drawn from the vertex M of one of the axes to the vertices of the other axis, the angle contained by these lines will be equal to the opposite angles ECF contained by the equal conjugate diameters.

PROP.

PROP. VI.

The square of the sum of the equal conjugate diameters of an ellipse is equal to the square of the sum of any other two conjugate diameters, together with the square of their difference.

LET the right lines GC, CF be equal conjugate diameters of an ellipse, and AB, BD any unequal conjugates, and AE their difference; the square of GF will be equal to the square of AD, together with the square of AE. FIG. 6.

For the squares of GC, CF taken twice are equal to the squares of AB, BD taken twice (by III. of this Book) but the squares of GC, CF taken twice are equal to the square of GF (4. 2.) and the squares of AB, BD taken twice are equal to the square of AD, together with the square of AE (10. 2.) therefore the square of GF is equal to the square of AD, together with the square of AE. *Q. E. D.*

COR. Hence it appears that the sum of equal conjugate diameters is greater, and the sum of the axes less than the sum of any other conjugate diameters; for the difference of the axes is greater than the difference of any other conjugate diameters, as is evident from Prop. VIII. Book II.

PROP.

PROP. VII.

The obtuse angle contained by equal conjugate diameters of an ellipse is the greatest, and the acute angle the least of all the angles which can be contained by any two conjugate diameters.

FIG. 7. LET AB be the transverse axis, and M*m* the conjugate axis of an ellipse: join AM, MB, the angle AMB is equal to the obtuse angle contained by the equal conjugate diameters (Cor. 2. V. of this Book) let CR, CT be any two unequal conjugate diameters, and draw from the vertex A of the transverse axis, AR an ordinate to CR, and meeting the ellipse in D, it will be parallel to CT, and join DB meeting CT in T; because AC, CB and AR, RD are equal, CR, TD will be parallel; therefore the angle ADB is equal to the angle RCT; but it is evident that the angles ADB, AMB are obtuse, being in a segment less than the semicircle described upon AB, as appears from (Prop. VIII. Book II.) if a circle be described through the points A, M, B, its segment AMB will be wholly within the ellipse: for the line BK touching the circle in B, and produced towards M, contains an acute angle KBA with the axis AB (because the alternate segment BEA is greater than a semicircle) therefore this line, and consequently the arch BM fall within the ellipse; but if it could meet the ellipse between the points B and M, and a line be drawn from this point ordinately applied to the conjugate axis M*m*, its other extremity would be in the periphery of the circle AMB, and therefore the ellipse and circle would meet in five points, contrary to (Cor. 3. LVI. Book I.) therefore the segment of the circle AMB is wholly within the ellipse, and therefore the angle AMB in this segment is greater than the angle ADB without the same segment: hence the Proposition is evident.

PROP.

PROP. VIII. PROBL. I.

Two conjugate diameters of an ellipse or hyperbola being given in position and magnitude, to find two other conjugate diameters which shall contain given angles.

If the given angles be right angles, the diameters required will be the axes, and the problem is solved by Prop. VII. Book II.

Case 1. WHEN the given diameters are the axes, let AB be the greater, and M*m* the lesser axis of the ellipse or hyperbola, and VZX, VZY the given angles; upon the greater axis AB describe the circle AFBG, so that the segments AFB, AGB may contain angles equal to the given angles VZX, VZY, and let FG be a diameter of this circle coinciding with the axis M*m*, and P be its center. FIG. 7, 8.

If the given obtuse angle VZX be equal to the angle contained by the lines AM, BM in the ellipse; the segment AFB of the circle meets the axis M*m* in its vertex M, in which case the diameters sought will be the equal conjugate diameters, by Cor. 2. Prop. V. and if the given angle VZX be greater than the angle AMB, the problem will be impossible by the preceding Prop. but if the given angle VZX be less than AMB, the segment AFB of the circle meets the axis M*m* produced in F: then draw through the point F a line parallel to MB, it will meet the circle somewhere between the points F and B, suppose in S; draw through the point S a line parallel to BC, it meets CM in O, which will be between the point F and the center C of the ellipse; then because, by a property of the circle, GO, OS, OF are proportional, GO will be to OF as the square of OS to the square of OF, that is (because the triangles OSF, CBM are similar) as the square of CB to the square of CM: in like manner in the hyperbola take a point O such, in the diameter FG produced, that GO be to FO as the square of CB to the square of CM, and in both sections from the center P, with the interval PO, describe

describe a circle meeting the axis AB in two points, and let Q be one of them; draw through the point Q a diameter of the circle AFBG meeting it in the points D, N; the point D, which is nearer to Q, will be in the ellipse or hyperbola, whose transverse axis is AB and M*m* its conjugate. For draw from the point D a right line perpendicular to AB, meeting it in H, and the circle again in L, and join LN; because the angle DLN is in a semicircle it is a right angle, and LN, HQ will be parallel; but the rectangle HLD is to the square of DH, as LH to HD, or as NQ to QD, that is, by construction, as GO to OF, or as the square of CB to the square CM, therefore the rectangle LHD or (by the circle) AHB is to the square of DH, as the square of AB to the square of M*m*: therefore the point D is in the ellipse or hyperbola, whose transverse axis is AB and conjugate axis M*m* (Cor. Prop. XXXIV. and XXXV. Book I.)

Let therefore the right lines AD, BD be inflected to the point D, and draw from the center of the section CR, CT, bisecting these lines, they will be the diameter sought. For because AR, RD and likewise AC, CB are equal, CR will be parallel to BD, and for the same reason CT will be parallel to AD; therefore CR, CT are conjugate diameters (XXVII. Book I.) and contain an angle RCT equal to ADB, and therefore contain angles (equal to those in the segments AFB, AGB of the circle, that is) equal to the given angles VZX, VZY. Because the circle described with the interval PO meets the axis AB in another point beside Q, it is manifest that in the same manner two other conjugate diameters may be found, which shall contain angles equal to the given angles VZX, VZY. *Q. E. I.*

If the circle AFBG be described upon the lesser axes, then, the diameter DQN being found as before, the point N, more remote from Q, will be in the ellipse or hyperbola, as is evident.

When the axes of the hyperbola are equal, describe upon either axis a circle, so that its segments shall contain angles equal to the given angles VZX, VZY, as before; and draw the right line DH

per-

perpendicular to the axis AB, meeting it in the point H, and touching the circle in D; because the square of DH is equal to the rectangle AHB, the point D will be in the hyperbola, and the diameters CR, CT, which bisect the right lines AD, BD, will be the diameters sought.

Case 2. When the given diameters are not the axes: because by Prop. VII. Book II. the axes may be found from two conjugate diameters given in position and magnitude, this case is reduced to the preceding; and when the position of the diameters sought is found, their vertices are found in the same manner as the vertices of the axes were found by Prop. VII. Book II.

Or this problem may be more generally solved, whatsoever conjugate diameters be given in position and magnitude, by describing upon one of the given diameters a circle, so that its segments shall contain angles equal to the given angles, and by finding the intersection of this circle with the ellipse, or with the hyperbola, when it is described about a transverse diameter; which may be done by a method similar to that, by which the intersection was found of the ellipse or hyperbola with the circle described upon its axis in the first case of this Prop.

PROP. IX. PROBL. II.

Two conjugate diameters of an ellipse or hyperbola being given in position and magnitude, and a right line given in position; to find the points in which this right line meets the section.

LET AB, M*m* be the given diameters, and C the center of the ellipse or hyperbola. Fig. 9, 10.

Case 1. First, if the line given in position be parallel to one of the conjugate diameters, and when the section is an ellipse, meets the other diameter between its vertices, or when the section is an

 hyper-

hyperbola, if it meets the tranſverſe diameter produced; then becauſe the line will be ordinately applied to this diameter, the points, in which it meets the ſection, may be found by Cor. XXXIV. and XXXV. Book I. and if it meets the conjugate diameter of the hyperbola, the points in which it meets the oppoſite hyperbolas may be found by Prop. XXXII. Book I.

Caſe 2. But if the line PQ given in poſition be not parallel to either of the diameters AB, M*m*: draw through the vertex of the diameter AB, the line AT parallel to the other diameter M*m*, it will touch the ſection (Cor. XXVII. Book I.) draw through the center a diameter parallel to PQ and meeting AT in H, and take a ſegment AK in the line AT (towards the point H in the hyperbola, and the contrary way in the ellipſe) ſo that the rectangle KAH may be equal to the ſquare of CM, and join CK; CH, CK will be conjugate diameters (Cor. Prop. LI. Book I.) draw therefore AE to the diameter CK and parallel to CH, it will be an ordinate applied to CK. Let therefore CE, CF, CK be proportional, and F will be the vertex of the diameter, which is in the line CK: in the ſame manner are found the vertices of the diameter, which is in the line CH: therefore this caſe is reduced to the firſt. But if in either caſe, the line PQ paſſes through the vertex of one of the conjugate diameters, and is parallel to the other, it will touch the ſection, as is evident.

Cor. Hence, two conjugate diameters being given in poſition and magnitude, a right line may be drawn which ſhall touch the ſection, and be parallel to a right line given in poſition.

PROP.

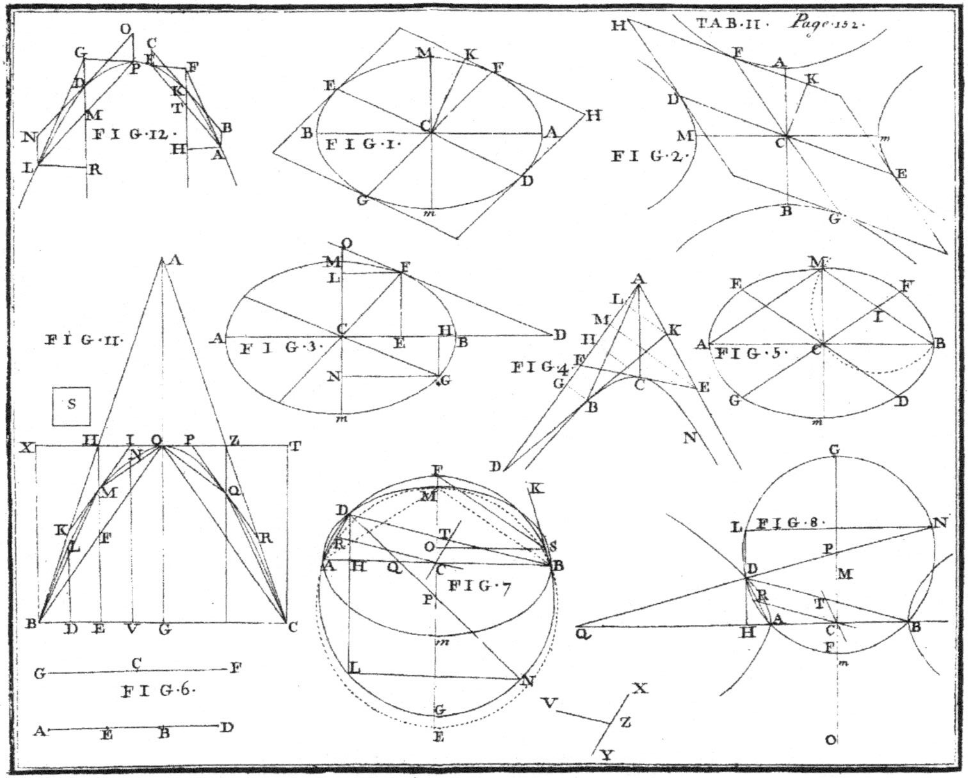
TAB. II. Page 152.
FIG. 1.
FIG. 2.
FIG. 3.
FIG. 4.
FIG. 5.
FIG. 6.
FIG. 7.
FIG. 8.
FIG. 11.
FIG. 12.

PROP. X. PROBL. III.

To deſcribe an ellipſe or oppoſite hyperbolas about a given parallelogram EFGH, ſo that the diameter of the ſection, parallel to the right lines EH, FG, ſhall be to its conjugate in a given ratio, viz. as TS to TQ. FIG. 11. 12.

DRAW the right line AB biſecting EH, FG, and the right line M*m* biſecting EF, HG; AB will be parallel to the lines EF, HG, and M*m* to the lines EH, FG; it is manifeſt that the interſection C of theſe lines will be the center of the ſection required, and that the diameters lying in the lines AB, M*m* will be conjugate; let the line AB meet EH in K, and the line M*m* meet EF in D and HG in V.

Caſe 1. Firſt, let the ſection to be deſcribed be an ellipſe, and as TS is to TQ, ſo let EK be to KL, and let KL be perpendicular to AB, from the center C, with the interval CL, deſcribe a circle meeting AB in the points A, B; and as TS is to TQ take CM or C*m* to CA; the ſection, whoſe conjugate diameters are M*m*, AB, is the ellipſe to be deſcribed. For by conſtruction, the ſquare of CM is to the ſquare of CA, as the ſquare of TS to the ſquare of TQ, and the ſquare of EK is in the ſame ratio to the ſquare of KL, or (by a property of the circle) to the rectangle BKA, and the right line EK is parallel to CM; therefore the point E will be in the ellipſe (Cor. XXXIV. Book I.) and becauſe EK, KH are equal, the point H will be in the ellipſe; and in like manner, becauſe ED, DF, and HV, VG are equal, the points F, G will be in the ellipſe. FIG. 11.

Caſe 2. Now let the oppoſite hyperbolas be the ſection to be deſcribed, through the points E, F, G, H; in this caſe the lines TS, TQ muſt not be to each other as EK to CK; therefore let EK be to KC in a ratio leſs than the ratio of TS to TQ, and let the line FIG. 12.

EK be to KL, as TS to TQ; KL will be leſs than KC; deſcribe a ſemicircle upon CK, and inſcribe in it the line KL, join CL, and take CA or CB equal to CL, and let CM or C*m* be to CA, as TS to TQ.

The oppoſite hyperbolas, of which AB may be a tranſverſe diameter, and M*m* its conjugate, will be the hyperbolas to be deſcribed; for by conſtruction, the ſquare of CM is to the ſquare of CA, as the ſquare of TS to the ſquare of TQ: or as the ſquare of EK (to the ſquare of KL, viz. the difference of the ſquares of CA, CK, that is) to the rectangle BKA (6. 2.) but EK is parallel to CM; therefore the point E will be in one of the oppoſite hyperbolas (Cor. XXXV. Book I.) and becauſe EK, KH are equal, the point H will be in the ſame hyperbola, and becauſe ED, DF, and HV, VG are equal, and theſe lines are parallel to the diameter AB, and meet its conjugate M*m* in D and V, the points F and G will be in the oppoſite hyperbola.

But if EK be to KC in a greater ratio than TS to TQ, let the ſquare of TS be to the ſquare of TQ as the ſquare of EK to the ſum of the ſquares of KC and of another line, ſuppoſe CA; then deſcribe from the center C, with the interval CA, a circle meeting CK in the points A, B, and as TS to TQ, ſo let CM or C*m* be to CA; the oppoſite hyperbolas, of which M*m* may be a tranſverſe diameter and AB its conjugate, will be the hyperbolas to be deſcribed: for the ſquare of CM is to the ſquare of CA as the ſquare of TS to the ſquare of TQ, or as the ſquare of EK to the ſum of the ſquares of CK and CA; therefore the hyperbolas will paſs through the points E, H, G, F, as is evident from Prop. XXXII. Book I. *Q. E. F.*

But if TS be to TQ as EK to KC, the problem in this caſe is impoſſible. For if the hyperbolas be deſcribed, whoſe conjugate diameters ſhall lye in the lines M*m*, AB, and ſhall be to each other as EK to KC, the point E will be in one of their aſymptotes.

Cor. 1. As it appears from the conditions of this problem that the diameters AB, M*m* are given in poſition and magnitude, it is evident

evident that one ellipse only, or two opposite hyperbolas only, can be described to answer the problem.

Cor. 2. Hence, an ellipse or opposite hyperbolas may be described about the given parallelogram EFGH, so that the ellipse, or one of the hyperbolas shall pass through a given point P. For draw the right lines AB, M*m* bisecting the opposite sides of the parallelogram, the conjugate diameters of the section to be described will be in these lines; draw through the point P a line parallel to AB, and meeting M*m* in R, and the side EH in O, and let RN be equal to PR: then describe through the points E, F, G, H, an ellipse, or the opposite hyperbolas, so that the square of the diameter in the line AB shall be to the square of the diameter in the line M*m* as the rectangle PON to the rectangle EOH; the point P will be in this ellipse, or in one of the hyperbolas, as is evident.

If the section to be described be an ellipse, the point P must be between the opposite sides, produced, of the parallelogram; but if it be an hyperbola, the point P must be within the parallelogram, or between the adjacent sides produced.

Cor. 3. Hence, an ellipse or opposite hyperbolas may be described, so that the given point C shall be their center, and the line MC*m* given in position a diameter, and likewise the parallels EF, PN bisected by MC*m* in the points D, R, shall be ordinately applied to that diameter. For take in the line M*m* the segment CV equal to CD, and through V draw GH parallel and equal to EF, and bisected in V: join HE, GF, and describe by the preceding Cor. the ellipse or opposite hyperbolas about the parallelogram EFGH, so that one of them may pass through the point P, this will be the section to be described: for the line M*m* will be a diameter, because it bisects the parallels EF, GH terminated by the ellipse or opposite hyperbolas, and the line PN parallel to EF, GH, will be ordinately applied to it; and because the equal ordinates EF, GH are equally distant from the point C, that point will be the center. If the given lines EF, PN be equal, and equally distant from the point C, several ellipses or hyperbolas may be described, which will answer the problem.

But

But if these lines EF, PN be equally distant from the point C, when they are unequal; or be at different distances from the same point when they are equal, the problem will be impossible. If the lesser line EF be at a greater distance from the point C than the greater line PN, the section to be described will be an ellipse; but if the contrary happens, opposite sections are to be described: and since in either case, it appears that the sections to be described must necessarily pass through six given points, from the conditions of the problem, it is evident that one ellipse only can be described, or two opposite hyperbolas only, which shall answer the problem, because two conic sections or opposite sections (Cor. 3. LVI. Book I.) can not meet each other in five, much less, in six points.

This Proposition and Corollaries, and likewise Prop. XIV. and XV. of this Book, are taken from Marq. de l'Hospital's Conic Sections.

PROP. XI.

FIG. 13, 14. Let a parallelogram KLNO be inscribed in an ellipse or opposite sections given in position; if the diameter AB, bisecting the opposite sides in the points R and S, has always a given ratio to its segment RS between those sides; the parallelogram KLNO will be always of the same magnitude.

LET C be the center of the section, CR, CS will be equal, as is evident: draw the diameter M*m* parallel to the sides LN, KO, it will be conjugate to AB; describe a parallelogram about these conjugates, it will be equiangular to the parallelogram KLNO, and therefore these parallelograms will be to each other in a ratio compounded of the ratios of their sides; by hypothesis, the ratio of AB to RS or KL is given; therefore the ratio of the square of CB is given (to the square of CR, and therefore) to the rectangle ARB (the difference of the squares of CB and CR) and consequently the ratio of the diameter M*m* to LN ordinately applied to AB

AB is given; therefore the ratio compounded of these given ratios of AB to KL, and of M*m* to LN is given; and therefore the ratio of the equiangular parallelograms is given: but the parallelogram AMB*m* is given in magnitude (Prop. I. of this Book) and therefore the parallelogram KLNO is always of the same magnitude. *Q. E. D.*

PROP. XII.

If the sides of a parallelogram EFGH touch an ellipse, or produced touch opposite hyperbolas given in position, and if the diameter AB of the section, which (if necessary produced) passes through the opposite angles E and G, has a given ratio to the segment EG between those angles; the parallelogram EFGH will be always of the same magnitude. FIG. 13, 14.

LET K, L, N, O be the points of contact, the right lines KN, LO joining the opposite points of contact will be diameters (Cor. 2. XXII. Book I.) and the four right lines joining these points will form a parallelogram KLNO; draw the diameter AB bisecting the parallels LN, KO in R and S, it will pass through the angles E and G (Cor. 1. XXVI. Book I.) in like manner the diameter M*m* bisecting KL, ON will pass through the other angles F and H, and will be conjugate to AB. Then because, by hypothesis, the ratio of CB to CG is given, the ratio of CB to CR is also given; therefore, as in the preceding, the ratio of the semidiameter CM to the ordinate LR, or to CT equal to LR, is given; therefore the ratio of CM to CF is given; and therefore if MP, FQ be drawn perpendicular to the diameter AB, the ratio of MP to FQ will be given; therefore the ratio compounded of these given ratios of AB to EG, and of MP to FQ is given, that is, the ratio of the rectangle under AB, MP to the rectangle under EG, FQ: but the rectangle AB, MP is given in magnitude (Prop. I. of this Book)

Book) therefore the rectangle EG, FQ, that is, the parallelogram EFGH is always of a given magnitude. *Q. E. D.*

PROP. XIII.

Fig. 15, 16. If the sides of a parallelogram EFGH touch an ellipse, or produced touch opposite hyperbolas; this parallelogram will be to a parallelogram described about any conjugate diameters of the section, as one of its sides HG is to that diameter of the section which is parallel to that side.

LET the diameter AB of the section be parallel to the sides HG, EF, and draw the diameter M*m* conjugate to AB, it will pass through the points in which the sides HG, EF touch the ellipse or opposite hyperbolas (Cor. XXVII. Book I.) let the parallelogram KLNO be described about these conjugate diameters, this and the parallelogram EFGH will be between the same parallels; therefore EFGH is to KLNO, as HG to (ON or) the diameter AB, which is parallel to HG: but KLNO is equal to the parallelogram described about any other conjugate diameters, whence the Prop. is evident.

Cor. 1. It is manifest from this Proposition, that the parallelogram described about any two conjugate diameters of an ellipse is the least of all the parallelograms which can be described about the same ellipse.

Cor. 2. Let AM and BM be joined, it is evident that the triangle AMB is the greatest that can be inscribed in this semi-ellipse, and consequently the parallelogram inscribed in the ellipse, by joining the vertices of the conjugate diameters, is the greatest that can be inscribed in the same ellipse; therefore the greatest parallelogram inscribed in an ellipse is half the least parallelogram circumscribed about the same ellipse. What has here been demonstrated concerning parallelograms inscribed in, or circumscribed about an ellipse

ellipse or about the hyperbolas, holds good of parallelograms inscribed in, or circumscribed about a circle.

PROP. XIV.

If from the center of an hyperbola, the segments CK, CL of an asymptote CN be taken, which shall be to each other as any other two parts CG, CH taken in the same asymptote CN: and if the right lines GF, HD, KB, LE be drawn parallel to the other asymptote CP, and meeting the hyperbola in the points F, D, B, E: and the semidiameters CF, CD, CB, CE be drawn; the hyperbolic sectors CBE, CDF will be equal. Fig. 17.

FOR draw the right lines BD, EF meeting the asymptotes in the points M, O and N, P, and because KB, HD, CO are parallel, MK will be to MB as HC to DO; therefore as MB, DO are equal, MK, HC will be equal: and in like manner, because LE, GF, CP are parallel, and NE, FP equal, NL, GC will be equal; therefore MK is to NL (as HC to GC, that is, by hypothesis, as LC to KC, or (on account of the hyperbola) as KB to LE: but the angles MKB, NLE are equal; therefore the triangles MKB, NLE are similar, and consequently the lines MB, NE, that is, BD, EF are parallel. Draw then the semidiameter bisecting the parallels BD, EF in the points A, Q; and because it will bisect all the lines terminated by the hyperbola, and parallel to BD, EF; it will bisect the area EBDF; therefore since the triangles CQF, CQE are equal, if from them the equal areas QADF, QABE and the equal triangles CDA, CAB be taken, the remainders, viz. the hyperbolic sectors CDF, CBE will be equal.

Cor. If the parts CG, CH, CK of the asymptote CN be in continued proportion, and GF, HD, KB be drawn, as in the proposition, and BF be joined; in the same manner it may be proved

X that

that the line BF is parallel to the tangent drawn through the point D; and thence that the sectors CDF, CDB are equal; and therefore if any number of parts CG, CH, CK, CL, &c. be in continued proportion, and the lines GF, HD, KB, LE, &c. be drawn parallel to the asymptote CP; the sectors CDF, CDB, CBE, &c. will be all equal to each other.

PROP. XV.

Fig. 17.

If the right lines DH, BK be drawn from the vertices of the semidiameters CD, CB to an asymptote CN, and parallel to the other asymptote CP; the hyperbolic sector CBD will be equal to the hyperbolic trapezium DHKB.

FOR the triangles CHD, CKB are equal, being the halves of equal parallelograms (Cor. 2. XLIII. Book I.) therefore if the common triangle CHR be taken from them, the triangle CDR will be equal to the trapezium HKBR, and if DBR be added to these equal spaces, the sector CBD will be equal to the hyperbolic trapezium DHKB.

Cor. Hence and from the preceding Corollary, if any number of parts CG, CH, CK, CL of the asymptote CN, be taken in continued proportion, and the lines GF, HD, KB, LE, &c. be drawn to the hyperbola, and parallel to the other asymptote CP, the hyperbolic trapeziums FGHD, DHKB, BKLE, will be all equal to each other.

PROP. XVI.

Being the Vth. of Archimede, concerning the Conoid and Spheroid.

The space contained by an ellipse is to the circle described upon the transverse axis of the ellipse, as the conjugate axis to the transverse axis.

Archi-

TAB 12 Page 160

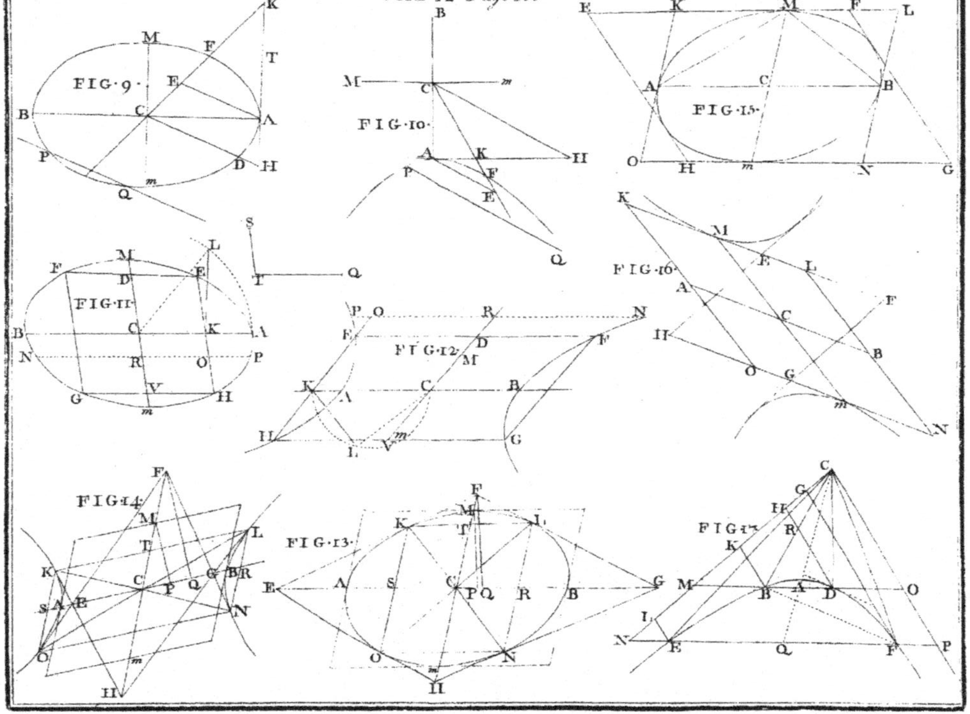

Archimede's Demonstration.

LET AC be the transverse axis and BD the conjugate axis of an ellipse, and let a circle be described upon the diameter AC; it is to be proved that the space contained by the ellipse will be to this circle, as BD to CA or EF: let the circle Z have to the circle AECF the same ratio as BD to EF; I say that the circle Z is equal to the ellipse. For if the circle Z be not equal to the space ABCD contained by the ellipse; first, if possible, let it be greater: then a polygon, having an even number of angles, may be inscribed in the circle Z greater than the space ABCD: let it be conceived to be inscribed, and inscribe in the circle AECF a polygon similar to that inscribed in the circle Z; and draw from its angles perpendiculars to the diameter AC, and let the points in which the perpendiculars cut the ellipse be joined by right lines: therefore there will be some polygon inscribed in the ellipse, which will be to a polygon inscribed in the circle AECF as BD to EF; for because the perpendiculars EH, KL are divided in the same ratio in B and M (Cor. 2. Prop. XXX. Book I.) it is evident, that the trapezium LE has the same ratio to LB, as HE to HB; therefore each of the other trapeziums in the circle will have to each of the other trapeziums in the ellipse the same ratio as EH to BH; and the triangles in the circle, at the points A, C, have the same ratio to those in the ellipse; therefore the whole polygon inscribed in the circle AECF will be to the whole polygon inscribed in the ellipse as EF to BD: but this same polygon is to the polygon inscribed in the circle Z in the same ratio; because the circles have this ratio; therefore the polygon inscribed in the circle Z is equal to the polygon inscribed in the ellipse, which is contrary to the hypothesis, for it was supposed greater than the whole space contained by the ellipse. Now let, if possible, the circle Z be less than the space ABCD, a polygon contained by an even number of sides, and greater than the circle Z, may be inscribed in the ellipse. Let it be inscribed, and draw from the angles perpendiculars to AC, and produce them to FIG. 18.

 the

the circumference of the circle: then again a polygon will be inſcribed in the circle AECF, which will have the ſame ratio to the polygon inſcribed in the ellipſe, as EF to BD, and if a polygon ſimilar to this be inſcribed in the circle Z, it may be demonſtrated, that the inſcribed polygon in the circle Z would be equal to the inſcribed polygon in the ellipſe, which is impoſſible; therefore the circle Z is not leſs than the ſpace contained by the ellipſe; and it has been proved that it is not greater, and therefore it is evident that the ſaid ſpace has the ſame ratio to the circle AECF as BD to EF. *Q. E. D.*

Cor. 1. Hence, it is manifeſt that the ſpace contained by an ellipſe is equal to a circle whoſe diameter is a mean proportional between the axes of the ellipſe.

Cor. 2. Hence, an ellipſe is to the rectangle under its axes as a circle to the ſquare of its diameter. For an ellipſe is equal to a circle whoſe diameter is a mean proportional between the axes of the ellipſe, and the rectangle under the axes is equal to the ſquare of the diameter of this circle; therefore the ellipſe is to that rectangle as that circle is to the ſquare of its diameter.

END OF THE FOURTH BOOK.

CONIC

CONIC SECTIONS.

BOOK THE FIFTH.

Of similar Sections, and of those Properties of the Conic Sections, which depend upon right lines harmonically divided; of circles which have the same curvature with the Sections: and of describing Sections, which shall pass through given points, and touch right lines given in position.

DEFI-

DEFINITION I.

IF two conjugate diameters of an ellipse or hyperbola be to each other as two conjugate diameters of another ellipse or hyperbola, and contain the same angles with them; these two ellipses or hyperbolas are said to be *Similar*.

PROPOSITION I.

FIG. 1, 2. Let two ellipses or hyperbolas be similar to each other, and so placed that they may have the common center C, and that two conjugate diameters AB, MN of one section may coincide with the two conjugate diameters *ab*, *mn* of the other section, to which they are proportional; I say that any other two conjugate diameters of the one section will coincide with two conjugate diameters of the other section, and that all the coinciding diameters will have the same ratio to each other.

Part 1. DRAW the conjugate semidiameters CD, CF of the exterior section, viz. of which AB, MN are diameters; then the semidiameters C*d*, C*f* of the interior section coinciding with CD, CF, will be conjugate.

For draw through the vertices M, *m* of the coinciding diameters two right lines touching the sections, they will be parallel to each other (by hypothesis, and Cor. XXVII. Book I.) let them meet the semidiameters CD, CF, when necessary produced, in the points L, O, and H, K; then by a property of parallel lines, the rectangles OML, K*m*H will be similar; therefore will be to each other as the squares of OM, K*m* or of CM, C*m*, that is, as the squares of CA, C*a* by hypothesis; but the rectangle OML is equal to the square of the semidiameter CA (LI. Book I.) therefore the rect-

rectangle K*m*H is equal to the square of the semidiameter C*a* of the interior section, and therefore C*d*, C*f* will be conjugate semidiameters of that section (by Cor. LI. Book I.)

Part 2. Now draw any two semidiameters CM, CD of the exterior section, and let C*m*, C*d* be the semidiameters of the interior section, coinciding with them; CM will be to C*m*, as CD to C*d*.

Join MD and draw CR bisecting MD in P, and let C*r* be the semidiameter of the interior section, and coinciding with CR; because, by the first part, the diameters which are conjugate to CR, C*r*, coincide, the ordinates applied to CR, C*r* will be parallel: draw *mp* an ordinate to C*r*, and let it be produced to meet CD in any point *d*; then because MD, *md* are parallel, and MP, PD equal; *mp*, *pd* will be equal: therefore because *mp* is ordinately applied to C*r* and the point *m* is in the interior section; the point *d* will be in the same section; and therefore the point *d* is the vertex of the semidiameter C*d*, and consequently because MD, *md* are parallel, CM will be to C*m* as CD to C*d*. *Q. E. D.*

Cor. 1. If two ellipses or hyperbolas are similar and similarly placed (that is, placed as in the proposition) and the right line TX terminated by the exterior section touches the interior one; it will be bisected in the point of contact: for let *m* be the point of contact, and draw through *m* the coinciding diameters CM, *cm*, and because the tangent TX is parallel to the diameter conjugate to C*m*, and the diameters conjugate to C*m*, CM coincide, by the first part of this Prop. the line TX will be an ordinate to the diameter CM, and therefore will be bisected in the point *m*.

Cor. 2. Or if the right line YZ terminated by the exterior section meets the interior one in two points *m*, *d*; the segments Y*m*, *d*Z, intercepted both ways between the sections, will be equal: for draw the diameter CR bisecting YZ in *p*; the diameter C*r* coinciding with CR will bisect *md* in the same point *p*; for because the diameters conjugate to CR, C*r*, coincide, the ordinates applied

to

to CR, C*r* are parallel; but the line Y*p*Z is ordinately applied to CR; therefore *mpd* is ordinately applied to C*r*; therefore *mp*, *pd* are equal, and consequently the segments Y*m*, *d*Z are equal.

Cor. 3. The same things supposed, if the line TX terminated by the exterior section touches the interior section in any point *m*, and a line be drawn through another point *d* in the interior section, parallel to TX, and meeting the exterior section in the points I, V; the rectangle I*d*V will be equal to the rectangle T*m*X or to the square of T*m*: for join the points *m*, *d*, and let this line meet the exterior section in the points Y and Z; because by the preceding Cor. the segments Y*m*, *d*Z are equal, the rectangles Y*d*Z, Y*m*Z will be equal; and therefore the rectangles I*d*V, T*m*X under the segments of the parallels which the line YZ meets, will be equal to each other (Cor. 1. Prop. XVIII. Book I.)

Cor. 4. The above construction supposed, if two lines MQ, TX terminated by the exterior section touch the interior section in the points *d* and *m*, they will be to each other as the semidiameters CF, CA, to which they are parallel: for through the point of contact *d* draw parallel to TX a line meeting the exterior section in I and V; the rectangle M*d*Q (by XXXI. and XL. Book I.) is to the rectangle I*d*V, that is, to the rectangle T*m*X (by preced. Cor.) as the square of CF to the square of CA; but the rectangle M*d*Q and T*m*X are the squares of M*d* and T*m*; therefore M*d*, T*m*, or the whole tangents MQ, TX are to each other as the semidiameters CF, DA, or as C*f*, C*a* coinciding with them.

Fig. 2. Cor. 5. The asymptotes of similar hyperbolas contain the same angles: for let two hyperbolas be similar and similarly placed, and the point C their common center, and let M, *m* be the vertices of the coinciding diameters, and A, *a* the vertices of their conjugate diameters, and because the sections are similar and similarly placed, the diameters CA, *ca* will coincide, and CM will be to C*m* as CA to C*a*; therefore the right lines joining the points A, M and *a*, *m* are parallel, and therefore if two right lines be drawn through the center C, one of them parallel to AM, *am*, and the other bisecting these

these lines, they will be asymptotes to both the hyperbolas (by Cor. 2. XXXVIII. Book I.) whence the Corollary is evident.

PROP. II.

Let EAD, *ead* be two parabolas, whose axes AB, *a*B coincide, and let them have the same parameter; I say that any other two diameters MN, *m*N coinciding will have equal parameters, and will contain with their ordinates equal angles. FIG. 3.

Part 1. LET F, *f* be the foci of the parabolas, and because they are equally distant from the vertices of their axes, by hypothesis and Def. of the focus, the distance F*f* between the foci is equal to the distance A*a* between the vertices of the axes, that is, to the distance M*m* between the vertices of the diameters MN, *m*N (by Cor. Prop. X. Book III.) therefore the lines MF, *mf* will be parallel and equal to each other (33. 1.) and consequently the diameters MN, *m*N coinciding, have equal parameters (Cor. 1. Prop. XXV. Book II.)

Part 2. Let the diameter MN be produced beyond the vertex to K, because MF, *mf* are parallel, the angles KMF, K*mf* will be equal; and therefore the right lines ML, *m*O bisecting these angles will be parallel (29. 1.) but they touch the parabolas (by Prop. XV. Book II.) therefore the coinciding diameters MN, *m*N contain equal angles with the tangents ML, *m*O and consequently with their ordinates. *Q. E. D.*

As circles differ only in the magnitude of their diameters, so parabolas differ only in the magnitude of their principal parameters, and therefore all parabolas, as well as all circles, are said to be similar.

COR. 1. If a right line TO terminated by the exterior parabola touches the interior in any point *m*, it will be bisected in that point;

for draw through the point of contact the diameter MN of the exterior parabola, and it is evident, from the second part of this Prop. that the line TO is ordinately applied to MN.

Cor. 2. If a right line ED terminated by the exterior parabola meets the interior in the points *e*, *d*; the segments E*e*, D*d* will be equal; for draw the diameter MN bisecting ED in P; and because by the second part of this Prop. the line *ed* will be an ordinate applied to the diameter *m*N, which coincides with MN; *e*P, P*d* will be equal, and therefore the segments E*e*, D*d* between the parabolas are equal.

Cor. 3. If a right line TO terminated by the exterior parabola touches the interior in any point *m*, and a right line be drawn from any point *d* in the interior parabola parallel to TO, meeting the exterior parabola in the points E, D; the rectangle E*d*D will be equal to the rectangle T*m*O or the square of T*m*; by joining the points *m*, *d*, it is proved by the preceding Cor. in the same manner as Cor. 3. of the preceding Prop.

Cor. 4. If two right lines TO, MQ terminated by the exterior parabola touch the interior in the point *m* and *d*; their squares will be to each other as the parameters of the diameters MN, OR, which pass through the points of contact; for these lines are ordinately applied to the diameters MN, OR (by Cor. 1.) and (by Cor. Prop. X. Book III.) the abscisses M*m*, O*d* are equal; therefore the squares of the ordinates T*m*, M*d* are to each other as the parameters of the diameters to which they are applied; whence the Corollary is evident.

Cor. 5. If two lines TO, MQ terminated by the exterior parabola touch the interior; they will cut off from the exterior parabola equal segments TMO, MOQ; for the diameters M*m*N, O*d*R being drawn through the points of contact; the abscisses M*m*, O*d* will be equal; therefore the parabolic segments TMO, MOQ will be equal (by Cor. 1. XIII. Book III.)

DEFI-

DEFINITIONS II, III.

II. IF a right line AD be so divided in the points C, B, that the whole line AD be to either of the extreme parts as the other extreme to the middle part CB; the line AD is said to be *harmonically divided.* FIG. 4.

COR. 1. It is evident that the middle part is less than either of the extreme parts.

COR. 2. Two extreme points A, D of a right line to be harmonically divided being given, and C one of the middle points; the fourth point B may be found, so that the given part AC shall be one of the extremes, viz. by dividing the segment CD in the point B, that the part CB, next to AC, be to BD as AC to AD.

COR. 3. Or the two middle points C, B being given, and A one of the extreme points; the other extreme D may be found; for draw the line AV from the point A, so that the extreme part AC may be to the middle part CB as AV to the segment VQ taken towards A; join BQ and draw VD parallel to QB, meeting ACB produced in D, and AD will be to BD as AV to QV, that is, by construction, as AC to CB.

COR. 4. It appears from the second Corollary, that, in harmonical division, two extreme points A, D, and the middle point C, and one extreme part AC, being all given, no other point besides B can be found to be a fourth point of that division; and it is evident from the third Corollary, that the middle points C, B and A, one of the two extreme points, and the middle part CB, all being given, no other point besides D can be found, which shall be a fourth part of that division.

Def. III. If a right line be harmonically divided in the points *a*, *c*, *b*, *d*, and in any manner four right lines V*a*, V*c*, V*b*, V*d* be drawn through the points of division, either parallel to each other, or meeting in the same point V; these lines are called *Harmonicals.*

LEMMA I.

Fig. 4. LET right lines V*a*, V*c*, V*b*, V*d* be harmonicals, meeting in the point V; if any right line parallel to one of them V*d*, meets the other three right lines in the points E, C, F; it will be bisected in the intermediate point C. Or if any four harmonicals meet in any manner a right line in the points A, C, B, D; this line will be cut harmonically in those points.

Part 1. Through the intermediate point C draw a parallel to the line *ad* (from which line the harmonicals are formed) and meeting them in the points A, C, B, D; it is evident that this line is divided in the same ratio as the line *ad*, that is, harmonically in the points A, C, B, D; therefore AD is to AC as BD to CB; but because VD, EC are parallel, VD is to EC (as AD to AC, that is, as BD to CB, that is) as the same line VD to CF (because VD, CF are parallel) therefore EC, CF are equal.

Part 2. Now let any right line meet four harmonicals in the points A, C, B, D, and if those harmonicals be parallel to each other, the Prop. is evident (2. 6.) but if they meet each other in a point V; draw through either of the middle points C the line ECF parallel to VD, which passes through the extreme point D more remote from the point C, and meeting the other two lines in the points E, F; by the first part EC, CF will be equal, and AD is to AC as VD to EC or CF, that is, as BD to CB; therefore AD is to AC as BD to CB. *Q. E. D.*

PROP.

PROP. III.

Let two right lines AB, AC, touching a conic section or opposite sections, meet each other, and draw BC joining the points of contact: if a right line be drawn through A the point of concourse of the tangents, meeting the section or opposite sections in the points E, K, and the line joining the points of contact in O; it will be cut harmonically in those points A, E, O, K. Fig. 5, 6.

Case 1. FIRST, let the line drawn through A not be a diameter, and draw two lines through the points E, K parallel to BC, joining the points of contact, meeting the tangents in D, G and H, M, and the section or sections in F and L: draw a diameter through A meeting the lines DG, BC, HM in the points N, P, Q, and because it bisects the line BC in P (Cor. 1. XXV. Book I.) it will bisect DG, HM in the points N and Q; and because EF, KL terminated by the section or sections are parallel to BC, they will be bisected in N and Q; therefore the segments DE, FG and the segments HK, LM will be equal, and consequently the rectangles DEG, HKM will be equal to the rectangles EDF, KHL.

Because DG, HM are parallel, HK will be to DE as KM to EG; therefore the rectangles HKM, DEG are similar, and therefore these rectangles, or the rectangles KHL, EDF are to each other as the squares of HK, DE, or as the squares of HA, DA; but these rectangles KHL, EDF are to each other as the squares of HB, BD (by Cor. 4. XVIII. Book I.) therefore the squares of HA, DA are to each other as the squares of HB, BD, and consequently HA is to DA as HB to BD (22. 6.) and from a property of parallel lines, KA is to EA as KO to OE; therefore the line AK is harmonically divided in the points A, E, O, K.

Case 2. When the lines touch the same section, and the line drawn through A is a diameter; let it meet BC in P, and the section or opposite sections in R, T; the lines drawn through these points Fig. 5.

points parallel to BC will be tangents; let them meet the tangent AB in S and V; by a property of parallel lines, VA is to SA as VT to SR, that is, as VB to BS (by Cor. 5. XVIII. Book I.) therefore, by a property of parallel lines, TA is to RA as TP to PR; and therefore the diameter drawn through A is cut harmonically in the points A, R, P, T. *Q. E. D.*

Cor. It appears from this demonstration that if the tangent VBS meets two parallel tangents VT, SR in the points V, S, and the line joining their points of contact in A; it will be harmonically divided in the points V, B, S, A, viz. in the point of contact, and in the points in which it meets the parallel tangents, and the right line joining their contacts.

PROP. IV.

If three right lines touch a conic section or opposite sections, each of them will be harmonically divided, viz. in the point of contact, and in the points in which it meets the other two tangents and the right line joining their points of contact.

IF two of the tangents be parallel, and the third meets the line joining their points of contact, the proposition is evident from the preceding Cor.

Fig. 7. But if the three lines AB, AC, MD, touching the section or opposite sections in the points B, C, K, meet each other in A, D, M, and the line BC joining two points of contact meets MD, if necessary produced, in H; HM will be to HD as MK to KD.

Through the point of intersection D of the tangents DK, DC, draw a right line parallel to the other tangent AB, and cutting the section, or either of the opposite sections in the points E, G, and meeting BC in F; the rectangle EDG will be equal to the square of DF (LIII. Book I.) but, by a property of parallel lines, the square of HM is to the square of HD as the square of MB to (the square of DF or) the rectangle EDG, that is (by Cor. 6. XVIII. Book

Book I.) as the ſquare of MK to the ſquare of KD; therefore HM is to HD as MK to KD. *Q. E. D.*

Cor. 1. Hence, if the right line BC joining the points of contact of the tangents MB, DC meets the other tangent MKD in H, and through its point of contact K a right line be drawn meeting BC in L, and parallel to AB, AC, if theſe lines be parallel, but if not, paſſing through A their point of concourſe; the line BC will be harmonically divided in the points H, B, L, C. For if MB, DC meet in A; join HA, and becauſe by this Prop. the tangent MKD is harmonically divided in the points H, M, K, D; the lines AH, AM, AK, AD will be harmonicals; therefore, by Part 2. Lemma preceding, the line HBC is harmonically divided in the points H, B, L, C.

But if the tangents MB, DC be parallel, the tangent MKD is harmonically divided in the points H, M, K, D, by Cor. to preceding Prop. therefore, becauſe MB, KL, DC are parallel, the line HBC is harmonically divided in the points H, B, L, C.

Cor. 2. Hence, if the two lines MB, DC be given in poſition, touching the ſection in the given points B, C, and any other point K in the ſection be given; the poſition of the tangent paſſing through K may be found. For draw through the point K a right line meeting BC in L, and parallel to MB, DC, if they be parallel, but if not, paſſing through their point of concourſe A; if LB, LC be equal, the line AKL is a diameter, and therefore the tangent paſſing through K will be parallel to BC. But if LB be leſs than LC, produce LB to H, ſo that HB be to HC as BL to LC (Cor. 3. to Def. II.) the right line joining the points H, K will be a tangent; for if not, draw a tangent through the point K, it neceſſarily meets ſomewhere LB produced towards H, ſuppoſe in N. Then by preceding Cor. the line HC, which by conſtruction is harmonically divided in H, B, L, C, will alſo be harmonically divided in the points N, B, L, C, contrary to Cor. 4. Def. II. therefore the line HK is a tangent.

PROP.

PROP. V.

FIG. 8, 9, 10, 11, 12. Let two points A, B be taken in a diameter DR of a conic ſection (on different ſides of the center, if DR be a conjugate diameter of an hyperbola, otherwiſe on the ſame ſide) ſo that the ſemidiameter CD of the ellipſe or hyperbola be a mean proportional between CA, CB, the diſtances of the points from the center C; but in the parabola ſo that the diſtances DA, DB of the points from the vertex D be equal to each other; and draw through theſe points A, B the right lines AT, BQ parallel to the ordinates applied to the diameter DR: if through either of the points A or B, ſuppoſe B, a right line be drawn cutting the ſection or oppoſite ſections in two points O and P; FO, FP, which touch the ſection or oppoſite ſections in thoſe points, will be either parallel to the line AT drawn through the other point A, or the point where they meet each other will be in that line.

IT is evident, if the ſecant drawn through the point B coincides with the diameter DR, that the tangents drawn through its vertices will be parallel to the line AT, becauſe they are parallel to the ordinates applied to the diameter DR.

But if the point B be within the ſection, or between the hyperbolas, and if the ſecant coincides with BQ, and meets the ſection or both ſections in Q and S, it is evident that the tangents drawn through Q and S will meet each other in the point A, becauſe in the ellipſe or hyperbola CB, CD, CA are proportional, and in the parabola DA, DB are equal.

Now let the ſecant OP be drawn, which ſhall not coincide with the diameter DR nor with the line BQ; then the tangents drawn through the points O and P will meet each other ſomewhere as in F: draw through the point F the diameter CEF biſecting OP in

G

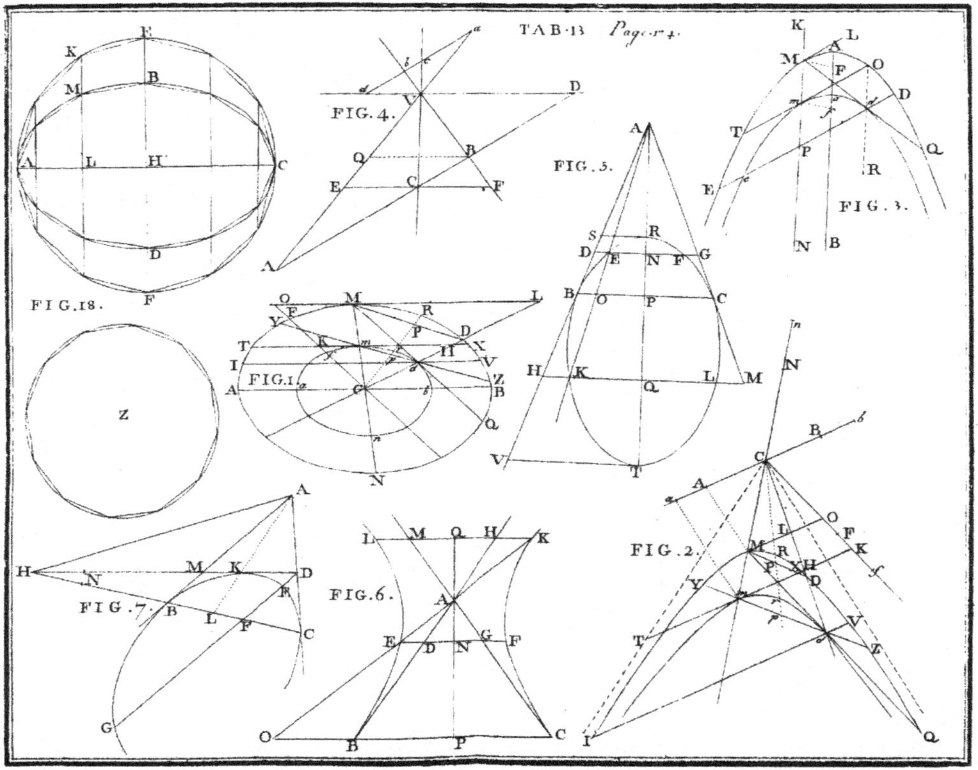
TAB·13
FIG.18.
FIG. 4.
FIG. 5.
FIG. 3.
FIG.1.
FIG.7.
FIG.6.
FIG.2.

G (XXVI. Book I.) and from the vertex D of the diameter DR between the points A, B, draw an ordinate applied to the diameter CE, and DH touching the ſection or conjugate ſection, and meeting the diameter CE in H; DH will be parallel to AT.

Becauſe, when the ſection is an ellipſe or hyperbola (by XLVIII. and XLIX. Book I.) each of the rectangles GCF, KCH is equal to the ſquare of the ſemidiameter CE, theſe rectangles will be equal to each other, and therefore CF is to CH, as CK to CG (or becauſe DK, OBG are parallel) as CD to CB, that is, by hypotheſis, as CA to CD; therefore if AF be joined, becauſe CF is to CH as CA to CD, AF will be parallel to DH or AT; and therefore the point F is in the line AT.

But when the ſection is a parabola, GE is equal to EF, and KE equal to EH (by XLVII. Book I.) therefore HF is equal to (GK or BD, becauſe DK, OBG are parallel, that is, by hypotheſis, to) DA; therefore becauſe HF, DA are equal and parallel, if AF be joined, it will be parallel to DH or AT; and therefore the point F is in the line AT. *Q. E. D.* Fig. 12.

Cor. 1. The ſame conſtruction remaining as above, if a right line be drawn through either of the points A or B, ſuppoſe B, meeting the ſection or ſections in two points M, L, and alſo AT drawn through the other point A, in F; it will be harmonically divided in the points F, M, B, L.

Firſt let the point B be within the ſection, and draw the diameter CE paſſing through F, and an ordinate through B applied to CE, meeting the ſection or ſections in the points O, P, the tangents drawn through O, P, meet ſomewhere in the diameter CEF (by Cor. 2. XXVI. Book I.) and they alſo meet in the line AT, by this Prop. and therefore meet in the point F, and conſequently the line drawn through B will be cut harmonically (by III. of this Book) in the points F, M, B, L. Fig. 8, 11.

But if the point B be without the ſection, and the line AT meets the ſection or ſections in the points T, N; ſince it appears from (Prop. LII. Book I.) that BT, BN joining theſe points are tan- Fig. 9, 10.

gents; the Prop. is evident by the III. of this Book. Supply the line LBM in Fig. 9, 10, 12.

Cor. 2. If a conic section be given in position, and any point B given, neither the center of the section nor in an asymptote of the hyperbola; any two right lines touching the section or opposite sections, and drawn through the extremities of the secant passing through the point B, meet each other in a right line as in AT, the position of which is determined by this Prop.

Cor. 3. Or if a conic section and any right line AT, not a diameter of the section, be given in position, and from any point F in AT two right lines be drawn, touching the section or opposite sections; the right line joining the points of contact will pass through a given point. For because the section and the line AT are given in position, the diameter CD will be given in position, whose ordinates are parallel to AT, and therefore its vertex D, and the point A in which point it meets AT will be given, and therefore in the diameter CD the point B will be given, placed with respect to the points A, D, as in the proposition: if the tangents FO, FP be drawn from any point F in AT, the line OP joining their points of contact will pass through the point B. For if not, a line might be drawn from the point B to one of the points of contact O, which meets again the section or opposite section in another point; then the tangent drawn through this point would meet the tangent FO in the line AT, by this Prop. that is, it would meet FO in F; and therefore three tangents would meet each other in the same point F, which is impossible (Cor. 3. XXVI. Book I.) therefore the Corollary is manifest.

PROP.

PROP. VI.

The same construction supposed as before, if through either of the points A or B, suppose B, BOP be drawn parallel to the asymptote CY, and meeting one of the opposite sections in O; a tangent drawn through the point O meets the line AT (drawn through the other point A) in the asymptote CY. Fig. 13. 14.

FOR since the tangent OF in Fig. 13. does not pass through the vertex D, viz. of the diameter DR between the points A, B, it will not be parallel to AT, but meets it produced towards the asymptote CY, and in Fig. 14. it is manifest that the tangent OF meets AT; then let OF meet AT in F; the point F will be in the asymptote CY.

For if not, from the point F another right line may be drawn (LII. Book I.) touching one of the opposite sections in some point as X; let OX be joined, and because the point F, from which point the tangents FO, FX are drawn, is in the line AT, and the point B is placed with respect to the points A, D, as in preceding Prop. the line OX will pass through the point B (by Cor. preceding) that is, will coincide with the line BOP; therefore BOP, which is parallel to the asymptote, meets the section or sections in two points O and X, which is impossible, and therefore the point F is in the asymptote CY. *Q. E. D.*

Cor. 1. If an hyperbola be given in position, and a point F given in its asymptote CY, and from the point F a right line be drawn meeting the hyperbola or opposite hyperbolas in the points N, T; tangents drawn through these points meet each other in a right line as OP, which will be given in position. From the point F draw a right line touching the hyperbola in O; because from the point F one tangent only can be drawn, the point O will be given, and therefore the line OP drawn parallel to CY will be given in posi-

 tion;

tion; let the tangents passing through the points N, T, meet each other in B, the point B will be in the line OP; for through B draw the diameter CD, and let D be its vertex, and let it meet NT in A; CB, CD, CA will be proportional, and AT an ordinate applied to the diameter CD; therefore the points B, D, A, and the line NAT are placed as in this Proposition: but if the point B be not in the line OP, through B a right line may be drawn parallel to CY, which shall meet the hyperbola in some point different from O, and the tangent drawn through that point will meet NAT in the asymptote CY, by the Proposition; therefore it meets it in F, and then two tangents meet each other in the same asymptote, which is impossible; therefore the point B, the concourse of the tangents drawn through the points N, T, is in the line OP.

Cor. 2. Or if an hyperbola and a right line OP parallel to its asymptote CY be given in position, and if from any point B in OP, two right lines be drawn touching the hyperbola or opposite hyperbolas in two points N, T; the right line joining those points will pass through a given point in the asymptote CY, to which OP is parallel. Through B let the diameter CD be drawn meeting NT in A, and through the point O, in which point OP meets the hyperbola, draw a tangent meeting the asymptote CY in F, and because CB, CD, CA are proportional, and the line NAT an ordinate applied to the diameter CD, and the tangent OF meets NAT in the asymptote CY, by this Prop. therefore the line NT joining the points of contact will pass through the point F: but because OP is given in position, the point O will be given, and consequently the tangent OF is given in position, but it meets CY given in position in F; therefore the point F is given.

Cor. 3. From any point F in the asymptote CY of an hyperbola, draw a right line touching the hyperbola in O, and draw OP through the point O parallel to CY: if through the point F a right line be drawn cutting the hyperbola or hyperbolas in the points N, T, and meeting OP in M, it will be divided harmonically in the points N, M, T, F.

For

For the tangents NB, TB drawn through the points N and T, meet each other in the right line OP, by Cor. 1. let the tangent OF meet NB, TB in E and G; becauſe, by Prop. IV. of this Book, the tangent OF is divided harmonically in the points F, G, O, E; BF, BG, BO, BE will be harmonicals: therefore the line NTF is harmonically divided in the points N, M, T, F (by Part 2. Lemma I.)

When the right lines NB, TB touch oppoſite hyperbolas, one of them as NB may be ſometimes parallel to the tangent OF; in this caſe the right line joining the points O, N will be a diameter; and becauſe the diameter BCA biſects TN in A, and OC, CN, and likewiſe TA, AN are equal, the right line joining the points O, T, will be parallel to BCA; therefore becauſe BOM is biſected in O (Cor. 2. LIV. Book I.) MT will be equal to TA or AN: and in like manner, becauſe OF, BN are parallel, MF will be equal to FN, therefore TF, FA are equal; therefore TA or MT is double of TF, and therefore MN is to FN, as MT to TF. Whence the Corollary is evident. FIG. 14.

PROP. VII.

The ſame conſtruction being ſuppoſed, if through either of the points A or B a right line parallel to an aſymptote of an hyperbola be drawn to the right line paſſing through the other point; it will be biſected by the hyperbola. FIG. 15.

FIRST, through the point A without the hyperbola, draw a line parallel to the aſymptote CY, and meeting the line BQ in P, and the hyperbola in O; becauſe CA, CD, CB are proportional, and the line BQ meets the hyperbola in the points Q, S, and is ordinately applied to the diameter CD, the lines joining the points A, Q and A, S will be tangents, and therefore the line AP is biſected in the point O (by Cor. 2. LIV. Book I.)

In the ſecond caſe, draw through the point B within the hyperbola a right line parallel to the aſymptote CY, and meeting AT in

F

F, and the hyperbola in M, BF will be bisected in M: for draw AOP parallel to CY, as in the preceding case, and through the point O a right line parallel to BQ, and meeting the diameter CDR in G, and the line BF in L; by a property of parallel lines, BP is equal to OL, and because AP is double of AO, (BP, that is) OL will be double of OG; therefore OL is bisected by the diameter CDR in G: but OL is an ordinate applied to that diameter, by construction, and its extremity O is in the hyperbola, therefore L the other extremity of it is in the same hyperbola; therefore the line OGL meets BF in the hyperbola, that is, in the point M, and therefore, because BP, MO, FA are parallel, and PO, OA equal, BM, MF will be equal.

PROP. VIII.

If two right lines, not parallel to each other, be inscribed in a conic section or opposite sections, and their extremities be joined by four right lines; their intersections, and the two intersections of the tangents which are drawn through the extremities of the two inscribed lines, will all four be in a right line.

FIG. 16. LET AC, BD be the two inscribed lines, which (produced if necessary) meet in E, and let AD, BC meet in F, and AB, DC in G, and the tangents drawn through A and C in H, and the other two tangents drawn through B, D in K; the four points F, G, H, K are in a right line.

For draw through the point E, the diameter NO of the section, and (in the ellipse and hyperbola) take NP a third proportional to the distance of the point E from the center N, and the semidiameter NO, placed in the order mentioned in Prop. V. of this Book; but in the parabola let OP be equal to the distance of the point P from the vertex O, and in both cases draw through the point P the right line PQ parallel to the ordinates applied to the diameter NO; the concourse of the tangents AH, CH, and of the tangents BK, DK

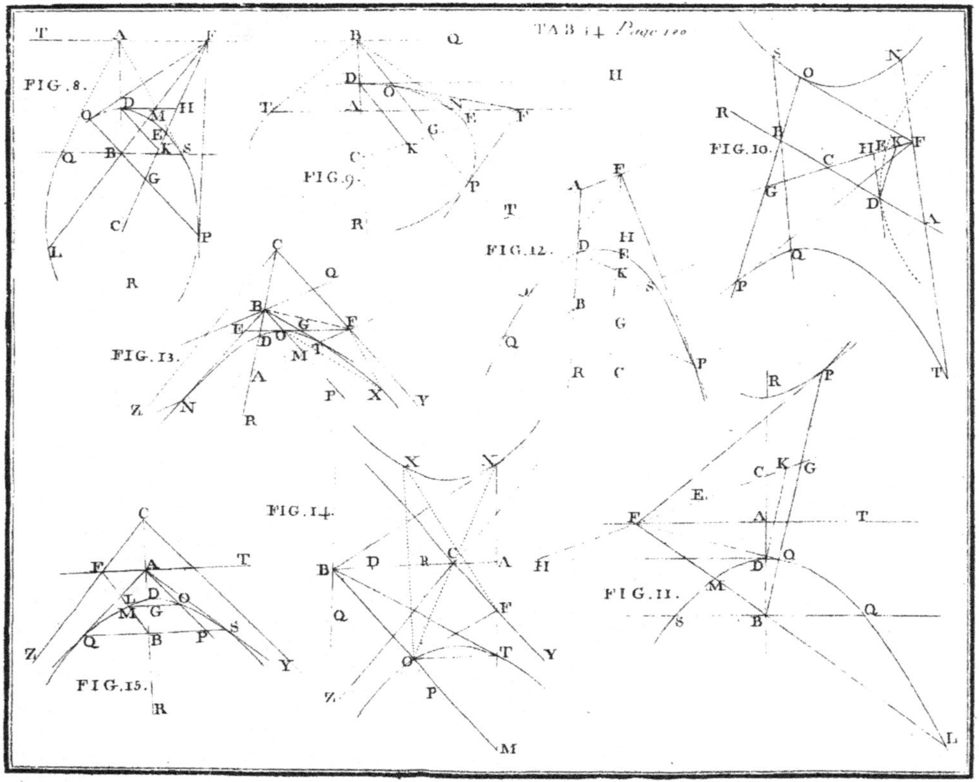

TAB 14 Page 120
FIG. 8.
FIG. 9.
FIG. 10.
FIG. 11.
FIG. 12.
FIG. 13.
FIG. 14.
FIG. 15.

DK, will be in the line PQ (by Prop. V. of this Book) let PQ meet AC, BD (if neceſſary produced) in the points L, M; then the line AC will be harmonically divided in the points A, L, C, E, and likewiſe the line BD in the points B, M, D, E (by Cor. 1. Prop. V. of this Book) let the line joining the points C, B meet PQ in F, and if the line joining the points A, D does not paſs through F, join DF meeting AC in X, and ſince B, M, D, E are points of harmonical diviſion, if FE be joined, FE, FD, FM, FB are harmonicals; therefore the line AC meeting theſe lines in the points X, L, C, E will be harmonically divided in thoſe points (by Part 2. Lemma I.) contrary to Cor. 4. Def. II. for it is divided harmonically in the points A, L, C. E; therefore AD paſſes through the point F: join AB meeting PQ in G, it may be ſhewn in the ſame manner that the line joining the points C, D paſſes through G; therefore the four points H, K, F, G are in a right line, viz. PQ. *Q. E. D.*

Cor. 1. But if the two lines AB, CD joining the extremities of the two inſcribed lines be parallel to each other, they will be parallel to PQ, in which line the tangents drawn through the extremities of the inſcribed lines meet; for if one of them AB meets PQ in any point G: then it may be ſhewn as before that the line DG paſſes through the point C, and therefore CD would meet AB in G, contrary to the hypotheſis; therefore the Corollary is manifeſt.

Cor. 2. If two right lines parallel to each other be inſcribed in a ſection or ſections; tangents drawn through the extremities of theſe parallel lines meet in the diameter biſecting them (Cor. 2. Prop. XXVI. Book I.) and it is manifeſt that the right lines joining the extremities of the parallels meet in the ſame diameter.

PROP.

PROP. IX.

FIG. 17, 18. If a right line cutting a section or opposite sections in the points P, Q, meets the directrix DX of the section in B, and three lines be drawn to the points P, Q, B, from the focus F nearer to DX, the line FB drawn to the directrix will bisect the angle QFP contained by the other two lines, when the points P, Q are in the opposite sections; but it will bisect the angle KFP adjacent to the angle QFP, when the points P, Q are in the same section.

FOR draw from the points Q, P, the perpendiculars QM, PN to the directrix, and QB will be to PB, as QM to PN, that is, as QF to PF (by Cor. 3. Prop. XI. Book II.) therefore when the points Q, P are in the opposite sections, the line FB bisects the angle QFP (3. 6.) but when the points are in the same section, from the point P draw a right line parallel to QF, and meeting FB in R, and QF will be to PR, as QB to PB, or QF to PF, as before; therefore PR, PF are equal, and therefore the angle BFP is equal to the angle PRF, or to the alternate angle BFK; therefore the line FB bisects the angle KFP adjacent to the angle QFP. *Q. E. D.*

FIG. 18. COR. 1. Hence, three points P, Q, V, being given in a conic section, and its focus F; the axis, passing through the focus, may be found both in position and magnitude; for join the points P, Q, and draw FP, FQ, and the line FR bisecting the angle adjacent to the angle PFQ, and meeting QB produced in B; the point B will be in the directrix: and in like manner, two other points Q, V being joined, find another point in the directrix, and draw the directrix DX: then draw from the focus and from the point P in the section the perpendiculars FD, PN to the directrix; and let FD be divided in the point A, so that FA be to AD, as PF to PN; the point A will be the vertex of the axis (XI. Book II.) and if FA, AD be equal, the section will be a parabola; but if they be unequal,

equal, take SF to SD, as FA to AD, and S will be the other vertex of the axis; and if FA be greater than AD, the point S must be taken on the same side of the focus where the point A lyes; but if FA be less than AD, on the contrary side; and in the first case the section will be an hyperbola, and in the other case an ellipse.

Cor. 2. Hence also, the directrix DX being given in position, and the focus F nearer to this directrix, and the point E in the section being given; a right line may be drawn to touch the section in the point E: draw through the points E, F, the right line EFO, and FB perpendicular to this line, meeting the directrix in B; the line joining the points E, B will be a tangent; for if not, let EB meet the section again in T, and produce the right line joining the points T, F to Y; by this Prop. the line FB bisects the angle EFY, which is adjacent to the angle ETF; therefore the angle YFB is equal to the angle EFB, that is, to the angle OFB, which is absurd; in like manner, Fig. 17. it may be demonstrated, that the line EB does not meet the section or opposite section again: therefore EB is a tangent. But if EFO be perpendicular to the directrix, the right line drawn through the point E, and parallel to the directrix, will be a tangent, as is evident. Fig. 18.

Cor. 3. Hence, if a right line EB touching the section meets its directrix DX in B, and EF be drawn from the point of contact E to the focus nearer to DX; the right line joining the points F, B will be perpendicular to EF; for if not, draw a perpendicular meeting the directrix in N, the right line joining the points E, N, will likewise be a tangent (by the preceding Cor.) which is absurd.

Cor. 4. Hence, if any right line EG, joining the points of contact of two tangents BE, BG, passes through the focus F; the right line FB drawn from the focus to the point of concourse of the tangents, will be perpendicular to the line joining the points of contact. For let DX be the directrix nearer to the focus F, and A the vertex of the transverse axis, between F and the line DX; then because the points F, A, D are placed, and DX drawn as in Prop. V. and EG passes through the focus F; the point B, the concourse of the tangents, will be in the directrix DX; therefore FB is perpendicular to EG, by the preceding Cor.

PROP. X.

If two right lines HP, HQ touch a conic section or opposite sections; the right line HF passing through the point of concourse of the tangents and focus F, will bisect the angle QFP contained by the right lines drawn from the points of contact to the same focus, when the lines HP, HQ touch the same section; but when they touch the opposite sections, the angle KFP, which is adjacent to the angle QFP, will be bisected by the line HF.

Fig. 17, 18. LET the right line DX be the directrix of the section nearer to the focus F, which the right line QP joining the points of contact meets in B, and which the axis meets in D; and let A be the vertex of the axis between the points F, D; through H the point of concourse of the tangents, draw a diameter meeting QP in L, and let V be its vertex, and let the line HF meet the section in E; and because in the ellipse or hyperbola, of which C is the center, CF, CA, CD, and likewise CH, CV, CL are proportional, and because FA, AD in the parabola, and likewise LV, VH are equal, and the lines DX, LP are parallel to the ordinates applied to the diameters DF, HL, and the line HF passes through as well the point F as the point H; according as this line may meet the section again in G, or only in E (when it is parallel to one of the asymptotes) the tangent drawn through E meets either the tangent drawn through G, or the asymptote parallel to HF, as well in the line DX as in the line QP (by Prop. V. and VI. of this Book) therefore the tangent drawn through E meets the directrix in B; therefore the line FB will be perpendicular to EFG (by Cor. 3. preceding) consequently the angles EFB, GFB are equal; but (QF being produced to K) the angle PFB (Fig. 18.) will be equal to the angle KFB (by the preceding Prop.) therefore the angle PFG is equal to the angle EFK, and consequently to the angle QFG: and in like manner, in Fig. 17. because the angles PFB, QFB are equal, the angle PFG is equal to the angle QFE, and consequently to the angle KFG, and therefore the Proposition is evident. *Q. E. D.*

PROP.

PROP. XI.

If two conic sections touch each other in one point (that is, if they have a common tangent in the same point) they cannot meet in three other points. Fig. 19.

LET two sections meet each other in the point A, and let the right line GA touch both sections in this point; these sections cannot meet in three other points besides the point A.

For if possible, let them meet each other in three other points B, C, D, and join these points by three lines, and first let none of them be parallel to GA; produce one of these lines BC that it may meet GA in G, and draw from the common point D a right line parallel to the common tangent GA, meeting GBC in N, and the sections in the points O, P; then each of the rectangles DNO, DNP will be to the square of GA, as the rectangle CNB to the rectangle BGC (by Cor. 3. XVIII. Book I.) therefore the rectangles DNO, DNP are equal, and consequently the points P and O coincide; therefore the sections meet each other in five points, contrary to Cor. 3. LVI. Book I. Or if the line DN touches one of the sections in D, and meets the other again in O, it may be demonstrated in the same manner that the square of DN is equal to the rectangle DNO, which is absurd. Or if the line DN touches both sections in D, because it is parallel to the tangent GA, the right line joining the points D, A will be a diameter of both sections, and therefore if from the common point B an ordinate be applied to the common diameter DA, the other extremity of this ordinate will be common to both sections, and consequently the sections meet each other in five points, which is impossible.

But if CD, joining two points common to both sections, be parallel to the common tangent GA, then a line drawn through the point of contact A, bisecting CD in L, would be a diameter common to both sections; if therefore there be drawn from the other common point B an ordinate applied to the diameter AL, the other extremity of this ordinate will be in both sections, and therefore these sections meet in five points, which is impossible, and therefore the Prop. is evident.

PROP. XII.

If two conic sections meet each other in two points; they will not meet in any other point.

Fig. 20. LET two conic sections touch each other in the points A, B, and first let the common tangents drawn through A and B meet each other in G, and draw through the point of concourse of the tangents, the right line GF bisecting the right line AB joining the points of contact, it will be a diameter to both sections (XXVI. Book I.) then if these sections have another common point D, draw DL an ordinate applied to the common diameter GF, the other extremity L of this ordinate will be in both sections, and therefore these sections touching each other in the point A, would meet in three other points B, D, L, contrary to what has been shewn in the preceding Prop. Or if the point C in the diameter GF be common to both sections, then the right line drawn through C and parallel to BA, would touch both sections in the point C (Cor. 8. XXV. Book I.) let this tangent meet the tangent GB in E; then the right line EH drawn through the point E, and bisecting the line which joins the points B, C, will be a common diameter to both sections, and therefore if there be drawn from the other common point A an ordinate applied to this diameter, the other extremity of it will be in both sections, which is contrary to the preceding Prop. as before: therefore there is no other point common to both sections, but the two points of contact A, B.

In the second place, if the common tangents drawn through A and B be parallel; the right line joining the points A, B will be a diameter of both sections; if they could have any other common point

point besides the two points of contact; then, from this common point, an ordinate being drawn to their common diameter, it may be shewn as before, that the sections would have a fourth common point, contrary to what has been proved in the preceding; and therefore in all cases, two sections which touch each other in two points do not meet each other in any other point. *Q. E. D.*

PROP. XIII.

If a circle touches a conic section, or opposite sections in two points; the right line joining the points of contact will be an ordinate applied to the axis of the section.

And if the ordinate be applied to the transverse axis of an ellipse or hyperbola; the circle falls wholly within the section; but if the ordinate be applied to the conjugate axis of an ellipse or hyperbola; the circle falls wholly without the ellipse, or without both the hyperbolas.

IF the common tangents be parallel, when the circle touches the ellipse or hyperbola in two points, the right line joining their points of contact will be a diameter of the section, and likewise of the circle; and because this diameter, by a property of the circle, is perpendicular to the tangents, it will be an axis of the section, as is evident from (Cor. 2. VII. Book II.) and if it be the transverse axis, the whole circle falls without the ellipse or both the hyperbolas; but if it be the conjugate axis of the ellipse, the whole circle falls within the ellipse, by Prop. VIII. Book II.

Part 1. But if the circle HRET touches the section or opposite sections in the points T, R, and the common tangents drawn through these points meet each other in S, the right line TR joining these points will be an ordinate applied to the axis of the section. For draw through the point of concourse S of the tangents a right line bisecting TR in O; then because (by the circle) the tangents ST, SR are equal, the line SO will be perpendicular to TR; but the line SO is a diameter of the section (XXVI. Book I.) and this diameter is the axis; for the line TR is ordinately applied to FIG. 21, 22.

to it at right angles; therefore the first part of the Proposition is evident.

Part 2. If the circle HTER touches a conic section in the points T, R, and the line TR joining these points be ordinately applied to an axis AB, which is the transverse in the ellipse and hyperbola; the whole circle will be within the section; and the center P of the circle will be in the axis AB, as is evident; let B be the vertex of the axis nearer to the point P, and let the circle meet the axis AB in E on the same side of the center P, where B is; PE will be less than PB.

For draw through the vertex B of the axis the tangent BL, and through the point R a right line touching the circle and section, and meeting the tangent BL in L; and because, in the ellipse or hyperbola, the conjugate axis, which is parallel to the tangent BL, is less than the diameter parallel to RL, BL will be less than RL (Cor. XXXI. and Cor. 1. XL. Book I.) and in the parabola BL is less than RL (by Cor. 1. III. Book II. and Cor. 3. III. Book III.) then join PL, BR, and because RL touches the circle, PRL will be a right angle; but PBL is a right angle; therefore, because of the common hypothenuse, the squares of PB, BL together, are equal to the squares of PR, RL together: but the square of BL is less than the square of RL, therefore the square of PB is greater than the square of PR, and consequently the line PB is greater than the line PR, that is, than PE; therefore the arch TER of the circle meets the axis within the section, and consequently that whole arch is within the section, otherwise the circle would meet the section in another point besides the two points of contact (contrary to what has been proved by the preceding) and as it is manifest that the circle meets the axis AB again in H within the section, the whole arch THR, and consequently the whole circle will be within the section: and in the same manner it may be demonstrated, that the circle falls wholly without the ellipse, when the right line TR joining the points of contact is an ordinate applied to the conjugate axis.

Because the circle touching the opposite hyperbolas is within the angle

angle contained by the common tangents, it is manifest it falls wholly without both the hyperbolas. *Q. E. D.*

Cor. 1. If any right line TS touches a conic section, and at the point of contact a perpendicular be erected to the tangent, and meeting the transverse axis of the section in P, PT will be the least line that can be drawn from the point P to the section: or if it meets the conjugate axis of the hyperbola or ellipse in the point Q, the line QT in the hyperbola will be the least, and in the ellipse the greatest of all the lines which can be drawn from the same point Q to the section, viz. on the same side of the axis. For draw from the point T a right line ordinately applied to the transverse axis, meeting the section again in R, and let the tangent TS meet this axis in S; SR joining these points (Cor. 2. XXVI. Book I.) will be a tangent, and equal to ST, and PR being drawn will be equal to PT: therefore the triangles PTS, PRS are equiangular, and consequently the angle PRS is a right angle; and therefore a circle described from the center P passing through the points T and R, touches the lines ST, SR, and consequently the section in T and R, and therefore that whole circle falls within the section, by this Prop. and therefore the radius PT of this circle is the least line which can be drawn from the point P to the section, viz. to the same part of the axis. It may be shewn in the same manner, when a perpendicular to the tangent TS meets the conjugate axis of the hyperbola or ellipse in Q, that the circle described with the radius QT falls wholly without the hyperbolas or ellipse. Wherefore the Corollary is manifest.

Cor. 2. If a right line TR terminated by an ellipse or hyperbola be an ordinate applied to its axis AB, and a circle passing through the points T and R, touches the section in one of the points T, it will likewise touch it in the other point R; for P the center of the circle will be in the axis AB of the section; draw through the point T the common tangent TS meeting the axis AB in S, PTS (from a property of the circle) will be a right angle, and the right line joining the points S, R touches the section; PR being joined, it may be shewn, as in the preceding Cor. that PRS is a right angle; there-

therefore the circle touches the line RS, and consequently the section in the point R.

PROP. XIV.

FIG. 21, 22. **If two circles touch an ellipse or hyperbola in the same point T, and touch again the same ellipse or opposite hyperbolas in the points R, N; the rectangle under the diameters of the circles will be equal to the square of that diameter of the section which is conjugate to the diameter passing through the point of contact T, and if a circle touches a parabola in two points, the square of its diameter will be equal to the rectangle under the parameter of the axis, and the parameter of the diameter passing through the point of contact.**

Part 1. LET AB, M*m* be the axes of an ellipse or hyperbola, and C the center; the right lines TR, TN joining the points of contact will be perpendicular to the axes, by Part 1. of the preceding Prop. through the point T, in which both the circles touch the section, draw a common tangent meeting the axes in the points G, S, and at the point T to this tangent erect a perpendicular meeting the axes in P and Q; the points P and Q will be the centers of the circles (1. and 19. 3.) and because the diameters of the section, which pass through the points of contact, are equal to each other, their conjugates will be equal to each other, as is evident from (Prop. III. and IV. Book IV.) let FD be the diameter of the section conjugate to that which passes through the point of contact T. Because the rectangled triangles TQG, TSP are equiangular, QT will be to GT, as TS to TP; and therefore the rectangle QTP is equal to the rectangle GTS, that is, to the square of CD, by Prop. LI. Book I. Whence the first part of the Prop. is evident.

Part 2. Let the circle HE touch the parabola in the points T, R, the right line TR joining these points will be perpendicular to the axis,

axis, as before; and therefore the parameters of the diameters, which pass through the points of contact, are equal (Cor. 3. XXV. Book II.) draw through the point T the common tangent meeting the axis in S, and erect at the point T a perpendicular to this tangent, meeting the axis in P, the point P will be the center of the circle HE as before; let TR meet the axis of the parabola in O, and because the rectangled triangles OPT, SPT are equiangular, OP will be to PT, as PT to PS; therefore the square of PT is equal to the rectangle OPS, that is, to the rectangle under half the parameter of the axis, and half the parameter of the diameter, which passes through the point of contact T (Prop. XXV. Book II.) therefore the second part of the Prop. is evident.

Cor. 1. The same things being premised, as in the first case of this Prop. the semidiameter PT of the circle touching the ellipse or hyperbola in two points, and falling within the section, will be to CD the semidiameter of the section conjugate to that which passes through the point of contact, as the conjugate semiaxis CM to the transverse semiaxis CB; for draw from the center the line CK perpendicular to the tangent GT: the rectangle under the lines PT, CK will be equal to the square of CM (by XXII. Book II.) and the rectangle under the lines CD, CK is equal to the rectangle MCB under the semiaxis, as is evident from (Prop. I. Book IV.) therefore PT is to CD, as the square of CM to the rectangle MCB, that is, as CM to CB: therefore, by this Prop. CD will be to QT (the semidiameter of the circle touching the ellipse or opposite hyperbola in two points, and falling without the sections) in the same ratio, as CM to CB.

Cor. 2. Hence it appears that the diameters of these circles will be equal to each other, if they touch (as in the Proposition) equilateral hyperbolas, viz. whose axes are equal to each other.

PROP. XV.

If from the vertex of the tranſverſe axis of an ellipſe or hyperbola, or from the vertex of the axis of a parabola, a line be taken on the axis equal to its parameter, and a circle be deſcribed about it as a diameter; it will fall wholly within the ſection: but if from the vertex of the conjugate axis of the ellipſe a line be taken equal to its parameter; a circle deſcribed about it as a diameter will be wholly without the ſection.

Fig. 23, 24, 25, 26. LET AB be the tranſverſe axis of the ellipſe or hyperbola, or the axis of the parabola, and take on the axis the line AC equal to its parameter; the circle deſcribed about the diameter AC falls wholly within the ſection: or if AB be the conjugate axis of the ellipſe, the whole circle falls without the ſection.

For draw AD perpendicular to the axis and equal to AC, and join CD, and in the ellipſe or hyperbola, let BLD be joined, but draw DL in the parabola parallel to the axis; and through any point E in the circle draw EF parallel to AD, meeting the ſection in G, and the lines AB, CD, LD in the points F, K, H; then becauſe AC, AD are equal, FC, FK will be equal; and by a property of the circle, the ſquare of EF is equal to the rectangle AFC, that is, to AFK; but the ſquare of GF is equal to the rectangle AFH (by Prop. I. IV. V. Book II.) but KF is leſs than FH, if AB be not the conjugate axis of the ellipſe, in which caſe KF is greater than FH; therefore in the firſt caſe, the rectangle AFK is leſs than AFH, and therefore the ſquare of EF will be leſs than the ſquare of GF, and the line EF leſs than FG, and conſequently the whole circle is within the ſection: in like manner it may be ſhewn in the other caſe that the whole circle is without the ſection. *Q. E. D.*

Cor.

Cor. 1. Hence, if from the vertex A of the transverse axis of the ellipse or hyperbola, a line AP be taken on the axis within the section, and not greater than half the parameter AC of this axis; AP will be the least of all the right lines which can be drawn from the point P to the section: but if a line AP be taken from the vertex of the conjugate axis of the ellipse, not less than half the parameter of the conjugate axis, PA will be the greatest of all the lines which can be drawn from the same point P to the section. For, in the first case, because the circle described with the radius PA does not fall without the circle AEC, it will be wholly within the section: and because this circle in the second case does not fall within the circle AEC, it will be wholly without the section, wherefore in both cases the Corollary is manifest.

Cor. 2. Hence, if a point P be given in the axis AB of a conic section (which in the ellipse or hyperbola is the transverse) and the distance of this point from the nearer vertex B of the axis is greater than the half of its parameter; a right line may be drawn from the point P, which shall be the least of all the lines which can be drawn from the same point to the section, viz. on the same side of the axis. Fig. 21, 22.

When the section is a parabola, take the line PO in the axis towards the vertex B equal to half the parameter of the axis, and when the section is an ellipse or hyperbola, take the line CO from the center C in the axis towards the vertex B, so that CO be to PO as the axis AB to its parameter, and draw through the point O in each section a line perpendicular to the axis, and meeting the section in T, the right line joining the points P, T shall be the line required. For draw through the point T a right line touching the section; the line PT will be perpendicular to this tangent, as is evident from (Part 1. Prop. XXV. and Prop. XXIV. Book II.) therefore PT is the least of all the lines which can be drawn from the point P to the section, viz. to the same side of the axis (by Cor. 1. Prop. XIII. of this Book).

And if a point Q be given in the conjugate axis of the hyperbola at any distance from the center, a line QT may be assigned in like manner, which ſhall be the leaſt of all the lines drawn from the ſame point to the ſame ſection: or if a point Q be given in the conjugate axis of the ellipſe, the diſtance of which point from the remoter vertex M of the axis is leſs than half its parameter; in the ſame manner, QT may be aſſigned, which ſhall be the greateſt of all the lines which can be drawn from the point Q to the ſection, viz. to the ſame ſide of the axis.

DEFINITION IV.

IF a circle touches a conic ſection in any point, ſo that no other circle can be deſcribed between this circle and the ſection, this circle is ſaid to have the ſame *curvature* with the ſection in the point of contact.

PROP. XVI.

Fig. 27, 28. From the point C in a conic ſection draw to its axis AH, which in an ellipſe or hyperbola is the tranſverſe axis, an ordinate meeting the ſection again in B, and from the point B, the diameter BK, and from the firſt point C an ordinate applied to this diameter meeting the ſection again in R: if a circle CRP be deſcribed touching the ſection in C, and paſſing through the point R, this circle will have the ſame curvature with the ſection in the point of contact C.

DRAW through the point C a common tangent, meeting the axis of the ſection in A, and through the point B a right line touching the ſection; this meets the other tangent in the point A of the axis (Cor. 2. XXVI. Book I.) and AC, AB will be equal, as is evident.

First,

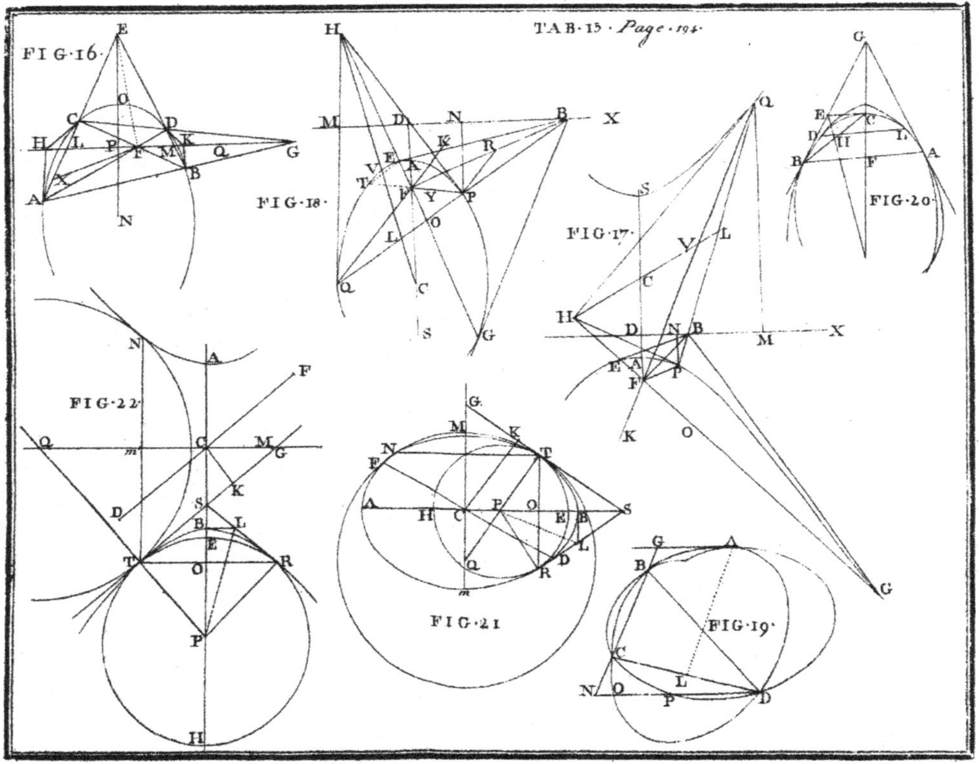
TAB·15 · Page ·194·
FIG·16·
FIG·18·
FIG·17·
FIG·20·
FIG·22·
FIG·21
FIG·19·

Firſt, I ſay that the circle CRP does not meet the ſection but in the points C and R; and that on one ſide of the line CR it is without, and on the other ſide within the ſection. For it cannot meet it in the point B, becauſe it would touch the ſection in that point (by Cor. 2. XIII. of this Book) and then could not meet the ſection in a third point R (by XII. of this Book) and by the ſame way of reaſoning it cannot meet the ellipſe in the other vertex of the diameter BK, for CK, being drawn, is an ordinate applied to the other axis. But if it be poſſible, let it meet the ſection in the points C, R and O, and draw a line through O parallel to BA, meeting the ſection in N, and AC in L; the rectangle OLN (XVIII. Book I.) will be to the ſquare of LC as the ſquare of BA to the ſquare of AC, that is, in a ratio of equality, and therefore, becauſe the point O is in the circle which LC touches in C, N alſo will be in the ſame circle; and therefore the circle CRP meets the ſection in the three points R, O, N, beſides the point of contact C, contrary to Prop. XI. of this Book, wherefore the circle CRP meets the ſection only in the two points C, R; and becauſe it does not touch the ſection in R (for then CR would be an ordinate applied to the axis (by the firſt part of Prop. XIII. of this Book, contrary to the hypotheſis) the arch of that circle on one ſide of the line CR will be wholly without, and on the other ſide wholly within the ſection.

Now if any other circle as COS be deſcribed leſs than CRP, FIG. 27. touching the circle CRP and the ſection in C, it falls within the ſection on both ſides of the point of contact C. For if that circle paſſes through B, it will touch the ſection in that point, and conſequently will be wholly within the ſection (Part 2. Prop. XIII. of this Book) then let it meet CB without the ſection in S; and becauſe the circle COS is wholly within the circle CRP, it neceſſarily muſt fall on one ſide of the point of contact C within the ſection: let the arch CM be within the ſection, and ſince the point S is without the ſection, the arch MS meets the ſection ſomewhere, ſuppoſe in the point O; draw a line through O parallel to BA, meeting the ſection in N and AC in L; then it may be demonſtrated as before,

fore, that the rectangle OLN is equal to the ſquare of the tangent LC, and conſequently the point N is in the circle COS; but that circle does not touch the ſection in N (XII. of this Book) becauſe it touches it in C, and meets it in the points N, O; therefore ſince the arch SN is on one ſide of the point N without the ſection, the arch NC will be on the other ſide of the point N within the ſection, and therefore that whole arch NC will be within the ſection: for otherwiſe the circle COS would meet the ſection in another point, beſides in the points C, O, N, contrary to Prop. XI. of this Book; therefore a circle leſs than CRP falls on both ſides of the point of contact C within the ſection.

Fig. 28. And if a circle CSOX greater than CRP be deſcribed, touching the circle CRP and the ſection in C, it falls on both ſides of the point of contact C without the ſection. For becauſe it is wholly without the circle CRP, it will neceſſarily be on one ſide of the point C without the ſection; therefore let the arch CXM be without the ſection, meeting the line CR produced in M; this circle CSOX meets the parabola or hyperbola (whoſe curves are indefinite) ſomewhere between the points M and C, ſuppoſe in O: and in the ellipſe, becauſe CB is an ordinate applied to the tranſverſe axis, CK will be an ordinate applied to the conjugate axis; therefore if the circle CSOX does not meet CK within the ſection, it falls wholly without the ellipſe (by Part 2. Prop. XIII. of this Book) let it then meet CK within the ellipſe in S; and ſince the point M is without the ellipſe, the arch MS meets the ellipſe ſomewhere as in O; draw therefore through the point O in each ſection a line parallel to BA, meeting the ſection in N and AC in L, it may be ſhewn as before, that the point N is in the circle CSOX, therefore becauſe that circle touches the ſection in C, it will not touch it again in the point N or O (by Prop. XII. of this Book) wherefore becauſe the point M is without the ſection, the arch MO will be without the ſection, and conſequently the arch OSN within the ſame, and therefore the arch NC on the other ſide of the point N will be without the ſection, and becauſe the circle CSOX does not meet the ſection but in the points C, N, O (XI. of this Book) the whole arch NC will

will be without the ſection; therefore the circle CSOX falls on both ſides of the point C without the ſection.

Then becauſe the circle CRP, touching the ſection in the point C. falls on one ſide of that point without, and on the other ſide within the ſection; and ſince every other circle touching the ſection in the point C falls on both ſides of the ſame point either within or without the ſection, it is manifeſt that no circle can paſs between the ſection and the circle CRP; and therefore that circle has the ſame curvature with the ſection in the point C, by the preceding Definition.

LEMMA II.

LET DS touch a circle in D, and through a point C within the circle, draw the line CX parallel to the tangent DS; if two right lines be drawn from the point D meeting the circle again in M, K, and the line CX in C, V; the rectangles MDC, KDV will be equal. FIG. 29.

Let the line CX meet the circle in Y, and join DY, MY; in the triangles DYM, DCY, the angle at D is common to both triangles, and the angle CYD is equal to the alternate angle YDS, that is, to the angle DMY in the alternate ſegment (32. 3.) and therefore the triangles DYM, DCY are equiangular; therefore MD is to DY, as DY to DC (4. 6.) and therefore the rectangle MDC is equal to the ſquare of DY: in the ſame way it may be ſhewn, KY being joined, that the rectangle KDV is equal to the ſquare of DY; therefore the Propoſition is evident.

PROP.

PROP. XVII.

If a tangent be drawn through any point in a conic section, and a circle be described touching this tangent in the same point, and cutting off from the diameter passing through this point a segment equal to its parameter; that circle will have the same curvature with the section in the point of contact.

Fig. 30. *Case* 1. LET AB be the axis of the section, and draw a tangent AD through its vertex A, BAD will be a right angle, and therefore if a circle ARC be described, touching the line AD and section in the point A, its center will be in the axis AB, and if that circle cuts off a segment AC of the axis equal to its parameter, it will be wholly within the section (or without the same, if AB be the conjugate axis of the ellipse) by Prop. XVI. of this Book, and therefore (when AB is not the conjugate axis) a circle, touching the section in A, and cutting off from the axis a segment less than AC, falls wholly within the section.

Then if a circle AHE cuts off from the axis of the conic section a segment AE greater than AC; let P be its center, and because the segment PA is greater than half the parameter of the axis, a line PT or P*t* on both sides of the axis AB may be drawn from the point P, which shall be less than PA (Cor. 2. Prop. XV. of this Book) therefore the circle AHE described about the diameter AE meets these lines produced without the section, suppose in the points M, L; and because the same circle meets the axis in E within the section, the arch ME necessarily meets the section somewhere between the points M, E, suppose in H: and in like manner the arch LE meets the section between the points L, E in *h*; and because that circle does not meet the section but in the points A, H, *h* (XI. of this Book) the arches AMH, AL*h* will be wholly without the section, that is, the circle AMEL falls on both sides of the point of contact

tact A, without the section: then because every other circle touching the section in the point A falls on both sides of that point either within the circle ARC, or without the section, it appears that no other circle can pass between the section and the circle ARC; therefore that circle, viz. which cuts off from the axis AB a segment AC equal to its parameter, has the same curvature with the section in the point A, by the preceding Definition. FIG. 29.

The demonstration is the same, when the point A is the vertex of the conjugate axis of the ellipse: for *lesser* reading *greater*, and for *within*, *without*, and *vice versa*.

Case 2. Let AB be the transverse axis of an ellipse or hyperbola, C the center, and through the point D, not the vertex of the axis, draw the line DS touching the section, and let the circle DMK be described touching this line in the same point D, and cutting off from the diameter CD a segment DM equal to its parameter; this circle will have the same curvature with the section in the point D. FIG. 29.

For from the point D let the ordinate DE be applied to the transverse axis, and meeting the section again in F, and draw through the point F the diameter FCG of the section, to which let DH be ordinately applied, meeting the section again in L, and the circle in K. Parallel to DS draw the diameter CQ meeting DL in V, and draw the semidiameter CR conjugate to CF, meeting DS in the point S, and let TD be an ordinate applied to this diameter; because the semidiameters CD, CF are equal, their conjugates CQ, CR will be equal; let PD be half the parameter MD of the diameter DCM, and the rectangle PDC will be equal to the square of the semidiameter CQ, or to the square of CR, that is, to the rectangle SCT, by the XLVIII. and XLIX. Book I. or to the rectangle VDH (because CS, VD and CH, TD are parallel) then because the rectangles PDC, VDH are equal, their doubles, viz. the rectangles MDC, LDV will be equal; but the rectangle KDV is likewise equal to the rectangle MDC (by the preceding Lemma) and therefore LD and KD are equal, consequently the points L, K coincide; therefore because the circle DKM touching the section in D, passes through the other extremity L of the line drawn from the point D, and ordinately applied to the diameter FG, this circle will have the ſame curvature with the ſection in the point D, by the preceding Prop.

FIG. 31. Now let the ſection be a parabola, its axis AB, and draw through the point D, not the vertex of the axis, a right line DS touching the parabola, and draw the diameter DG; let a circle DMK be deſcribed touching the line DS in D, and cutting off from the diameter DG the ſegment DM equal to its parameter; that circle will have the ſame curvature with the parabola in the point D.

For draw from the point D an ordinate DE applied to the axis, meeting again the parabola in F, and through the point F draw the diameter FH, to which, from the point D, let DH be ordinately applied, meeting again the parabola in L, and the circle in K: from the vertex of the diameter FH, draw the ordinate FG, applied to the diameter DG; this ordinate will be parallel to the tangent DS, let it meet DH in C. Then becauſe the parameters of the diameters DG, FH are equal (Cor. 3. XXV. Book II.) and the abſciſſes DG, FH are equal (II. Book III.) the ordinates DH, FG will be equal; but becauſe the triangles DCG, FCH are ſimilar, and the ſides DG, FH equal, DC will be equal to CH, and conſequently DC will be a fourth part of the whole line DL; therefore the rectangle LDC will be equal to (the ſquare of DH or FG, that is, from a property of the parabola) to the rectangle MDG; but by the preceding Lemma, the rectangle KDC is equal to the rectangle MDG; and therefore as before, the circle DMK will paſs through the point L, and conſequently will have the ſame curvature with the parabola in the point D. *Q. E. D.*

COR. Hence, if a circle DBF touches a parabola in two points D, F, and a circle DMO be deſcribed, paſſing through the point of contact D, and having the ſame curvature with the parabola in the point D; the parameter of the axis of the parabola, the diameter of the circle DBF, the parameter of the diameter DG, and the diameter of the circle of curvature, will be in continued proportion: the line joining the points D, F, will be an ordinate applied

to

to the axis (XIII. of this Book) which let it meet in E, draw through the point D a right line touching the parabola, and meeting the axis in S; this line likewise touches both the circles; erect at the point D a perpendicular to DS, meeting the axis in X, XD will be the radius of the circle touching the parabola in the points D and F, and the segment DM of the diameter DG cut off by the circle of curvature, will be equal to the parameter of that diameter; therefore if from the center of this circle the perpendicular NP be drawn to DM, PD will be half of DM, and consequently equal to SX, and EX will be half the parameter of the axis (Prop. XXV. Book II.) but EX is to XD as XD to XS, and as XD is to XS, so PD or XS is to ND, because the triangles DXS, PDN are similar; therefore the whole lines, viz. the parameter of the axis of the parabola, the diameter of the circle DBF, the parameter of the diameter DG, and the diameter DO of the circle of curvature, will be in continued proportion.

PROP. XVIII.

If a circle DMO touching an ellipse or hyperbola in the point D, has the same curvature with the section in that point; its semidiameter ND will be to CQ the semidiameter of the section, conjugate to that which passes through the point D, as the square of CQ to the rectangle ACZ under the semiaxes. Fig. 19.

If the circle touches the section in the vertex of the axis, its diameter will be the parameter of the same axis, by Prop. XVI. of this Book: therefore the Proposition is manifest in this case.

But if the circle DMO touches the section in a point D, not the vertex of the axis; draw through the point D a common tangent DS, and from the center of the section the perpendicular C*a* to this tangent: let DM be the segment of the semidiameter CD cut off by the circle DMO, and draw from the center of that circle the perpendicular NP to the diameter CD; and PD will be half the parameter of the diameter CD; therefore the rectangle PDC

 is

is equal to the ſquare of the ſemidiameter CQ conjugate to CD; but becauſe the triangles NDP, DC*a* are ſimilar, ND will be to CD as PD to C*a*; and therefore the rectangle under the lines ND, C*a* is equal to (the rectangle PDC, that is, to) the ſquare of CQ; but ND is to CQ as (the rectangle under ND, C*a*, that is, as) the ſquare of CQ to the rectangle under the lines CQ, C*a*; but the rectangle CQ, C*a* is equal to the rectangle ACZ under the ſemiaxes (I. Book IV.) and therefore the ſemidiameter ND of the circle DMO, is to the ſemidiameter CQ as the ſquare of CQ to the rectangle ACZ under the ſemiaxes. *Q. E. D.*

Cor. Becauſe it is ſhewn in the demonſtration of this Prop. that the rectangle under the lines ND, C*a* is equal to the ſquare of CQ, it is evident that the radius ND of the circle of curvature, the ſemidiameter CQ of the ſection parallel to the common tangent DS, and the perpendicular C*a* drawn from the center of the ſection to the tangent, are in continued proportion.

PROP. XIX.

Fig. 32. **If a right line cutting a conic ſection in two points D, F, meets two tangents AB, AC in the points E, G, and the line BC joining their points of contact in H; the rectangle DEF will be to the rectangle FGD, as the ſquare of EH to the ſquare of GH.**

DRAW through the point E, where the ſecant meets one of the tangents AB, a line parallel to the other tangent AC, and meeting the right line joining their points of contact in M, and draw a line parallel to DF touching the ſection in L, and meeting the tangent AC in K; by Cor. Prop. LIII. Book I. the rectangle DEF is to the ſquare of EM as the ſquare of KL to the ſquare of KC, that is, as the rectangle FGD to the ſquare of GC (by Prop. XVIII. Book I.) therefore, by permutation, the rectangle DEF will be to the rectangle FGD as the ſquare of EM to the ſquare of GC,

TAB. 16. *Page 202*

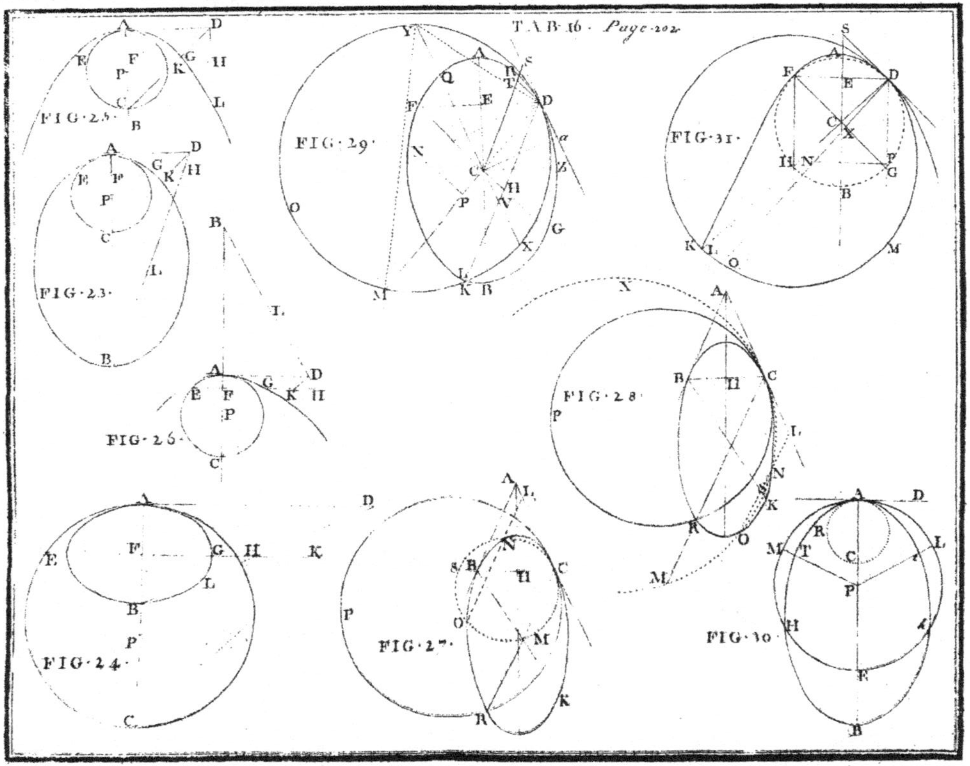

GC, that is, because the triangles HEM, HGC are similar, as the square of EH to the square of GH. *Q. E. D.*

If the tangents, which the secant meets, be parallel to each other, the Prop. is manifest, from Cor. 2. XVIII. Book I.

PROP. XX.

If two right lines DF, CA cutting a conic section in the points D, F and C, A, meet each other in G, and meet in the points E, H, a right line touching the section in B; the square of EB is to the square of BH in a ratio compounded of the ratio of the rectangle DEF to the rectangle CHA. and of the ratio of the rectangle AGC to the rectangle FGD. FIG. 33.

Case 1. DRAW through the point E, in which the secant DF meets the tangent, a line parallel to the other secant AC, which in the first place may meet the section in the points K, L; because these secants are parallel, the square of EB will be to the square of BH as the rectangle KEL to the rectangle CHA; but the ratio of the rectangle KEL to the rectangle CHA is compounded of the ratio of the rectangle KEL to the rectangle AGC, that is, of the ratio of the rectangle DEF to FGD, and of the ratio of the same AGC to CHA; therefore the ratio of the square of EB to the square of BH is compounded of the ratio of the rectangle DEF to the rectangle FGD, and of the ratio of the rectangle AGC to CHA, that is, of the ratio of DEF to CHA, and of the ratio of AGC to FGD.

Case 2. Now if any line be drawn parallel to DF, meeting the section in the points *d*, *f*, the line AC in *g*, and the tangent in *e*, through which point a line drawn parallel to AC would not meet the section; then from what has been demonstrated, the ratio of the square of *e*B to the square of BH is compounded of the ratio of the square of *e*B to the square of EB, and of the ratio of the square of EB to the square of BH; therefore it is compounded of the ratio

ratio of the rectangle *def* to DEF (Cor. 4. XVIII. Book I.) and of the ratio of the same rectangle DEF to CHA, and of the ratio of AGC to FGD, by the first case, that is, of the ratio of the rectangle *def* to CHA, and of the ratio of AgC to *fgd*, (because the rectangle DEF is both an antecedent and a consequent of the terms which compound the ratio, and therefore may be omitted, and the rectangles AgC, *fgd* may be substituted for the rectangles AGC, FGD, which have the same ratio to each other (by Prop. XVIII. Book I.) therefore in this case the Prop. is demonstrated.

The two preceding Propositions are in part demonstrated by Sir Isaac Newton, in Prop. XXIII. and XXIV. Book I. Math. Prin. Philos. Nat.

PROP. XXI. PROBL. I.

Five points being given in a conic section, to describe the section.

FIG. 34. LET the five points F, G, K, M, N be given in a conic section, and join four of these points by two right lines FG, MN, meeting each other in the point R, and draw through the fifth point K two lines KD, KH parallel to FG, MN, and meeting the lines MN, FG in E and Q; and take the points D, H in the lines KD, KH (if necessary produced) such that the rectangle KED be to MEN as GRF to MRN, and KQH to GQF as MRN to GRF; (but the points K, D, or K, H must be on the same or different sides of the points E or Q, according as the points M, N, or G, F are on the same or different sides of the points E or Q) it appears from Prop. XVIII. Book I. that the points D and H are in the conic section passing through the five points F, G, K, M, N; then draw the right line IL bisecting the parallels KD, FG terminated by the section, it will be a diameter of the section, and draw another diameter AB bisecting the parallels KH, MN in O and P; if these diameters are parallel to each other the section will be a parabola; and since four points are given in it, viz. K, M, N, G, the section may be described by the second case of Prop. IX. Book III.

III. But if the diameters meet each other, as in the point C, the section will be an ellipse or hyperbola, which may be described by Cor. 3. Prop. X. Book IV. for its center C is given, and the diameter AB is given in position, to which the parallels KH, MN are ordinately applied.

In this and the following problems, when the section to be described is an ellipse, it may sometimes be a circle, which is to be considered as a species of the ellipse.

PROP. XXII. PROBL. II.

To describe a conic section which shall pass through three given points, and touch two right lines given in position.

LET AV, AB be the lines given in position: and first let the section to be described touch these lines in the given points B, V, and pass through the third given point within the angle BAV. FIG. 35.

If the lines AB, AV be parallel, the section to be described will be an ellipse, and the line BV joining the points of contact will be a diameter, to which the line drawn through the third point given, parallel to AB, AV, will be an ordinate, and therefore in this case the section will be easily described, by Cor. Prop. XXXV. Book II.

Case 1. Now let the lines AB, AV meet in the point A, and let BV be bisected in P, and draw PA, it will be a diameter of the section to be described (XXVI. Book I.) then let the third given point O be in the line PA; if PO, OA are equal, describe a parabola (by XXXIII. Book II.) of which PA shall be a diameter, whose vertex is O, and to which BV shall be ordinately applied; the lines AB, AV touch this parabola in B and V, as is evident from Prop. XLVII. Book I. But if the point O be between P and A, and the lines PO, OA unequal, let the lesser line be produced to C, so that CP, CO, CA be proportional, and take CQ equal to CO; then (by Cor. XXXV. Book II.) describe an ellipse or hyperbola, of which let OQ be a diameter, and the line VB be ordinately applied to this diameter, the section touches the lines AB, AV in the points B and V, as appears from Prop. XLVIII. XLIX. Book

Book I. if PO be leſs than OA, the ſection will be in an ellipſe, but if greater, an hyperbola. If the given point be Q in the line PA, and not between P and A, take the point C between P and Q, ſo that CP, CQ, CA be proportional, and let CO be equal to CQ, then deſcribe an ellipſe, as before, to the diameter OQ, it touches the lines AB, AV in the points B and V, as is manifeſt.

If, the points B, V remaining the ſame, the third given point K be not in the diameter AP of the ſection: draw KE parallel to BV, meeting the tangent AB in L, and AP in F, and let FE be equal to KF, and take from the point L towards K the line LG, ſo that the ſquare of LG be equal to the rectangle KLE, join BG meeting AP in O; then, as before, deſcribe the ſection touching the lines AB, AV in B and V, and paſſing through O, it will paſs through the given point K; for if not, let the ſection meet the line KE in R, and take FT equal to RF, becauſe RF is parallel to BV, it will be ordinately applied to the diameter AP, and therefore the point T will be in the ſection, and KR, TE equal; draw through the point O a right line touching the ſection, and meeting the tangent AB in N, it will be parallel to LRT; therefore the rectangle RLT is equal to the ſquare of LG (by Prop. LIII. Book I.) that is, to the rectangle KLE, which is abſurd; therefore the ſection paſſes through the given point K.

FIG. 36. *Caſe* 2. Let the lines AB, AV be given in poſition, and the three given points K, E, F between theſe lines; a ſection is to be deſcribed, which ſhall touch theſe lines, and paſs through the three given points.

Let the lines which join the points K, F and K, E, meet AB, AV in the points D, G and L, N, and take in the line KF the point H ſuch that the ſquare of HD be to the ſquare of HG as the rectangle KDF to the rectangle FGK, and take in the line KE the point P, ſo that the ſquare of PL be to the ſquare of PN as the rectangle KLE to the rectangle ENK, and let the right line joining the points H, P be drawn; if this line meets in the points B and V, the ſides containing the angle BAV, within which the three given points are taken, deſcribe a conic ſection, by the preced-

ing

ing cafe, touching the lines AB, AV in B and V, and paffing through the point K, it will likewife pafs through the points F and E; for if this fection does not pafs through F, let it meet the line KF in X; then (by Prop. XIX. of this Book) the rectangle KDX will be to the rectangle XGK as the fquare of HD to the fquare of HG, that is, by conftruction, as the rectangle KDF to the rectangle FGK, which is abfurd; therefore the fection paffes through the point F: and in like manner it may be fhewn that it will pafs through the point E.

If the line KF, joining two of the three given points, be parallel to AV, one of the lines given in pofition; then the point H may be taken in this line on either fide of the point D, fo that the fquare of DH be equal to the rectangle KDF, as is evident from Prop. LIII. Book I.

When the fegment DK, FG, and confequently the rectangles KDF, FGK be unequal, it is manifeft that the point H may be taken either between the points K and F, or on the other fide of the leffer fegment DK, and in like manner the point P may be taken on either fide of the fegment LK; and likewife if the line drawn through the points E and F, meets the lines AB, AV, there may be found in it two points in the fame way as the points H, P were found, and fometimes twelve fuch lines as HP may be drawn, and as many of thefe lines as meet the fides of the angle BAV, fo many fections may be defcribed, which fhall anfwer the problem in this cafe.

Cafe 3. If two lines AB, AV be given in pofition, and a fection be required to be defcribed, which fhall touch the line AV in the given point V, and the line AB in another point, and pafs through the given points K, F; join KF, and take in this line a point H as in the preceding cafe, and draw VH meeting AB in B; then, as before, let a fection be defcribed touching AB, AV in the points B and V, and paffing through K, this fection will pafs through the point F, as in the preceding cafe. If the point H can be taken between, or not between the points K, F, two fections may be defcribed, which fhall anfwer the problem.

In this Propofition all the cafes are folved of Prop. XXIII. Book I. of Newton's Math. Prin.

PROP. XXIII. PROBL. III.

Four points being given in a conic ſection, and a right line given in poſition touching the ſection; to deſcribe the ſection.

Fig. 33. *Caſe* 1. LET A, F, D, C be the four points given in the ſection, and EH the line given in poſition touching the ſection: draw the lines AC, FD meeting the tangent in the points E and H; if theſe lines be parallel to each other, take a point B in the tangent between E and H, ſo that the ſquare of EB be to the ſquare of BH as the rectangle DEF to the rectangle CHA, B will be the point of contact, by Cor. 4. XVIII. Book I. and if the ſegment EB be greater than the ſegment BH, the point B may be taken on the other ſide of the point H; but if the lines AC, FD meet each other in G, then the point of contact B may be found, by Prop. XX. of this Book, which point, as before, may be taken either between the points E and H, or not between; therefore ſince five points A, F, D, B, C are given in a conic ſection, it may be deſcribed by (Prop. XXI. of this Book).

Fig. 37. *Caſe* 2. Let A, B, C, D be the given points in the ſection, and let GB touch it in B; join the four given points by right lines, forming a trapezium; firſt, let none of the ſides of this trapezium be parallel, and let the diagonals meet in E, and produce the ſides AD, BC to meet each other in F, ſo that EF joining theſe two points may meet GB in G, (which may be always done) then becauſe two lines AB, DC, not parallel to each other, are inſcribed in a conic ſection, and their extremities are joined by four lines meeting each other in the points E and F, and becauſe the tangent drawn through B meets EF in G, the tangent drawn through A meets EF in the ſame point G (by Prop. VIII. of this Book) therefore the line joining the points A, G touches the ſection required. Deſcribe then, by the firſt caſe of the preceding Prop. a ſection which ſhall touch the lines GA, GB in the points A, B, and ſhall paſs through the point C, it will alſo paſs through the point D; for if not, let it meet

meet BD in X, then the line joining the points A, X, would meet BC in F (by VIII. of this Book) which is absurd; therefore the section will pass through the point D.

Then, if the two sides AB, DC of the trapezium be parallel, the right line EF bisecting these parallels will be a diameter, and if the tangent GB meets EF, as in G, then the tangent drawn through A will meet EF in the same point G, in which case the section must be described as before: but if the tangent GB be parallel to the diameter EF, the tangent drawn through A will be parallel to the same diameter EF, and the section, to be described, an ellipse, in which the line AB will be the diameter conjugate to EF, and therefore if a line be drawn from the point D or C to AB, parallel to EF, it will be an ordinate applied to the diameter AB, and from hence (by Cor. 1. XXXI. Book I.) the vertices of the diameter EF may be found, and the ellipse described by Prop. XXXIV. Book II.

But if the right lines joining the four given points form a parallelogram the section will be an ellipse, and the lines which bisect the opposite sides of this parallelogram will be conjugate diameters, and their magnitudes are found, by Prop. XLVIII. Book I. *Q. E. F.*

PROP. XXIV. PROBL. IV.

Five right lines, which touch a conic section, being given in position; to find their points of contact, the section not being given in position.

LET the lines AB, BC, CD, DE, EA touch a conic section; their points of contact are to be found. Fig. 38.

Let ABCDE be the figure contained by the five tangents, and call AB the *first side*, BC the *second side*, and in like manner the other sides: and let FBCD be a quadrilateral contained by the first four sides, and draw the diagonals BD, FC meeting each other in M: now omitting the first side AB, let ICDE be a quadrilateral contained by the other four sides, and draw the diagonals ID, CE meeting each other in N; and draw MN; this line will pass

 through

through the points in which the second side BC and the fourth side DE touch the section.

For let G, H, K, L, O be the points of contact of the lines AB, BC, CD, DE, EA, and join GH, LK and GL, HK: then because the lines GH, LK are inscribed in a section, and B is the point of concourse of the lines which touch the section at the extremities of the line GH, and D the point of concourse of the lines which touch the section at the extremities of LK; the points B, D, and the point of intersection of GK, LH will be in a right line (by Prop. VIII. of this Book). Again, because GL, HK are inscribed in a section, and the lines which touch the section in G, L meet each other in F, and the lines touching the section in H, K meet in the point C; the points F, C, and the point of intersection of GK, LH are in a right line; but the same point of intersection has been shewn to be in the line BD; therefore it must be in the point of intersection of BD, FC, that is, in M; which point therefore will be in the line LH. Join LO, and because the lines OH, LK are inscribed in a section, and the tangents which touch the section in O, H meet in the point I, and the tangents which touch the section in L, K meet in the point D, the points I, D, and the point of intersection of OK, LH will be in a right line; and because OL, HK are inscribed in a section, it may be shewn in the same manner that the points E, C, and the point of intersection of OK, LH are in a right line; wherefore it will be in the point of intersection of the lines ID, EC, that is, in the point N: therefore N is in the line LH, and it has been shewn that M is in the same line; therefore MN passes through the points L, H, viz. of contact of BC, DE: and in like manner the points of contact of the other tangents may be found.

FIG. 39. COR. And likewise, if in a conic section five points A, B, C, D, E be given; the lines which touch the section in those points are easily found, the section not being given in position.

Let A be called the *first point*, B *the second*, and so on; join the four first points A, B, C, D by the right lines A B, B C, CD, DA, meeting each other in the points F, G: likewise join AC, BD; then because

becauſe the lines AC, BD are inſcribed in a ſection, (by Prop. VIII. of this Book) the points F, G, and the point of concourſe of the lines which touch the ſection in B, D will be a right line, that is, FG will paſs through the point of concourſe of theſe tangents. Omitting the point A, let BCDE be a quadrilateral contained by the other four ſides; and let the points H, K be the interſections of the ſides, and join CE; then becauſe EC, BD are inſcribed in a conic ſection, the points H, K of interſection of the ſides, and the point of concourſe of the tangents, which touch the ſection in B, D, will be in a right line, that is, HK will paſs through this point of concourſe of the tangents; but this ſame point has been ſhewn to be in the line FG; therefore it muſt be in L the point of interſection of the lines FG, HK; and therefore LB, LD touch the ſection in the points B, D: and in the ſame manner, the tangents may be found, which ſhall touch the ſection in the other points.

This Propoſition and Corollary are taken from the Conic Sections of Dr. Robert Simſon. See Cor. 2. and 3. Prop. XLVII. Book V. of the firſt edition.

THE END.

TAB·17 Page·212

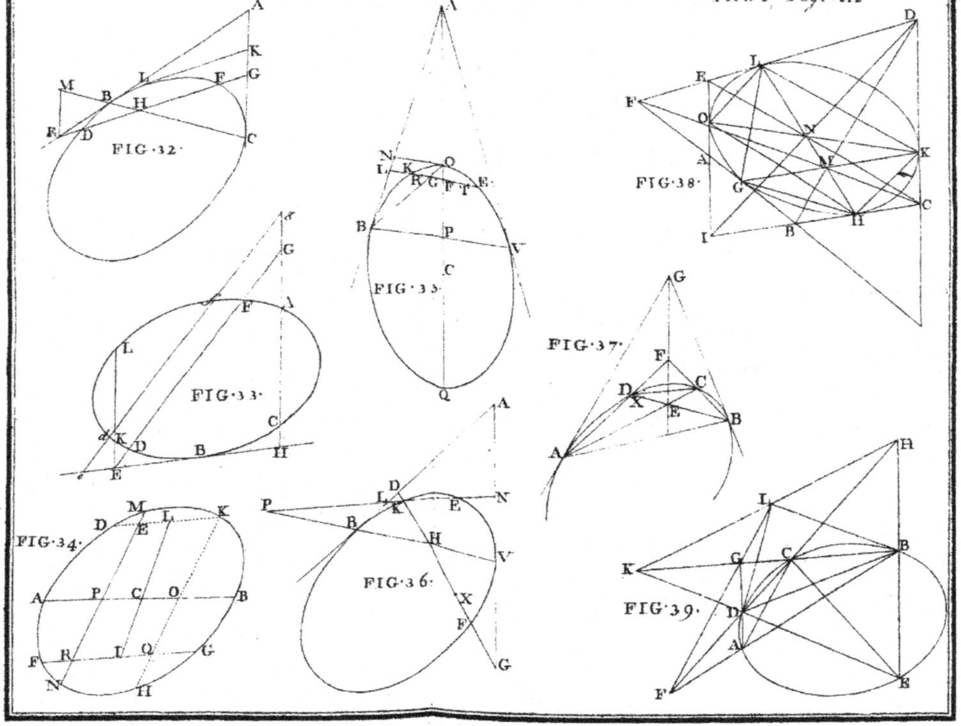

FIG·32· FIG·33· FIG·33· FIG·34· FIG·36· FIG·37· FIG·38· FIG·39·

Lately published by the same Author,

I.

PHILOSOPHICAL ESSAYS

On the Following Subjects.

I. On the Ascent of Vapours, the Formation of Clouds, Rain, and Dew, and on several other Phenomena of the Air and Water.

II. Observations and Conjectures on the Nature of the Aurora Borealis, and the Tails of Comets.

III, On the Principles of Mechanics.

The Second Edition Improved and Enlarged.

II.

The DUTY of OBEDIENCE to the LAWS, and of SUBMISSION to MAGISTRATES.

A SERMON

Occasioned by the late Disturbances in the North of Ireland, preached before the Judges of Assize in the Cathedral Church of Armagh, on Sunday April 12, 1772.

Printed for J. NOURSE, opposite Catharine Street in the Strand.

Where may be had, lately published,

I. CYCLOMATHESIS: Or an Easy Introduction to the several Branches of the Mathematics; being principally designed for the Instruction of Young Students, before they enter upon the more abstruse and difficult Parts thereof. In Ten Volumes.

By W. EMERSON.

VOL. I. Containing, 1. A General Introduction to the Cyclomathesis. 2. A Treatise of Arithmetic.

VOL. II. 1. The Doctrine of Proportion, Arithmetical and Geometrical. 2. The Elements of Geometry.

Note, the above Two Volumes may be bound in One.

VOL. III. 1. The Elements of Trigonometry, Plain and Spherical. 2. A Table of Natural Sines and Tangents. 3. A Table of Logarithmic Sines and Tangents. 4. A Table of Logarithms from 1 to 10,000.

VOL. IV. A Treatise of Algebra, in Two Books.

VOL.

VOL. V. 1. The Arithmetic of Infinites, and the Differential Method. 2. Elements of the Conic Sections. 3. The Nature and Properties of Curve Lines.

VOL. VI. The Elements of Optics and Perfpective.

VOL. VII. 1. Mechanics, or the Doctrine of Motion. 2. The Projection of the Sphere, Orthographic, Stereographic, and Gnomonical. 3. The Laws of Centripetal and Centrifugal Force.

VOL. VIII. A Syftem of Aftronomy.

VOL. IX. 1. The Mathematical Principles of Geography. 2. The Theory of Navigation, Spherical and Spheroidical. 3. Dialling: Or the Art of Drawing Dials on all Sorts of Planes.

VOL. X. 1. The Doctrine of Combinations, Permutations, and Compofitions of Quantities. 2. Chronology: Or the Art of Reckoning Time. With a Chronological Table. 3. Calculation, Libration, and Menfuration: Or the Art of Reckoning, Weighing, and Meafuring. 4. The Art of Surveying, or Meafuring of Land.

N. B. Any of the above Volumes may be had feparate, to accommodate fuch Gentlemen that do not chufe to purchafe the whole.

II. The METHOD of INCREMENTS, wherein the Principles are demonftrated, and the Practice thereof fhewn in the Solution of Problems. By W. Emerfon, Quarto.

III. NAVIGATION: Or the Art of Sailing upon the Sea, containing a Demonftration of the Fundamental Principles of this Art, together with all the Practical Rules of computing a Ship's Way, both by Plain Sailing, Mercator, and Middle Latitude, founded upon the foregoing Principles, with many other ufeful Things hereto belonging. To which are added, feveral neceffary Tables. The Second Edition, corrected and enlarged, 12mo. By W. Emerfon.

IV. The ELEMENTS of EUCLID, viz. the firft fix Books, together with the eleventh and twelfth. The Errors by which Theon and others have long ago vitiated thefe Books are corrected, and fome of Euclid's Demonftrations are reftored. Alfo the Book of Euclid's Data, in like Manner corrected. By Robert Simfon, M. D. Emeritus Profeffor of Mathematics in the Univerfity of Glafgow. To this fourth Edition are alfo annexed, Elements of Plain and Spherical Trigonometry. 8vo.

N. B. The Elements of Trigonometry may be had feparate to complete the former Edition.

Made in the USA
Monee, IL
13 September 2026

62976817R00148